Bird versus Bulldozer

BIRD VERSUS BULLDOZER

A QUARTER-CENTURY CONSERVATION BATTLE IN A BIODIVERSITY HOTSPOT

AUDREY L. MAYER

Yale
UNIVERSITY PRESS
New Haven and London

Yale University Press books may be purchased in quantity for
educational, business, or promotional use. For information, please e-mail
sales.press@yale.edu (U.S. office) or sales@yaleup.co.uk (U.K. office).

Maps throughout this book were created using ArcGIS® software by
Esri. ArcGIS® and ArcMap™ are the intellectual property of Esri and
are used herein under license. Copyright © Esri. All rights reserved.

Set in Adobe Garamond and Gothic types
by Tseng Information Systems, Inc.
Printed in the United States of America.

Library of Congress Control Number: 2020939830
ISBN 978-0-300-24790-9 (hardcover : alk. paper)

A catalogue record for this book is available from the British Library.

This paper meets the requirements of ANSI/NISO Z39.48-1992
(Permanence of Paper).

10 9 8 7 6 5 4 3 2 1

For Bill and Luukas

Contents

Preface

The title of this book is brazen for a tiny, gray songbird that few people know. The bird isn't the bald eagle, peregrine falcon, or whooping crane, but the California gnatcatcher. The gnatcatcher doesn't have the widespread notoriety of other threatened and endangered species that have stood in the way of economic growth, such as the northern spotted owl. This lack of infamy may be due to an innovative state-level conservation policy, the Natural Community Conservation Planning Act, which places environmental protection on an equal footing with economic growth in rapidly urbanizing Southern California. The fact that most Americans are blissfully unaware of the California gnatcatcher suggests that the NCCP policy may offer valuable lessons for harmonizing nature conservation and economic activity.

The California gnatcatcher eats insects and spiders, and weighs about the same as a U.S. quarter. Its song sounds as if a kitten is plaintively whining through the high end of a harmonica. The species is nonmigratory but capable of searching for new breeding habitat when necessary, since gnatcatchers evolved in a region where fire regularly requires them to move. California gnatcatchers breed exclusively in a habitat called coastal sage scrub, which looks and smells exactly how it sounds—a collection of waist-high bushes with pleasant aromatic qualities, found along the California coast from Santa Barbara down through Baja California, Mexico. The habitat is one of my favorite places in which to do fieldwork, particularly since my end-of-day field clothes smell far better than they do after walking around in other habitats. Unfortunately, California gnatcatchers and people both favor these coastal areas for housing—a contest that gnatcatchers usually lost.

Since California achieved statehood in 1850, the human population of Southern California has grown from thousands to over twenty million people. Those millions produced food and built homes in one of the most biologically diverse landscapes in the world—a hotspot of species richness and uniqueness. As farms and neighborhoods expanded from the coastline into the valleys and foothills, almost 90 percent of the California gnatcatcher's habitat was destroyed, and their numbers in the United States declined precipitously. The loss of California gnatcatchers and their habitat prompted the U.S. Fish and Wildlife Service to list them as threatened under the U.S. Endangered Species Act in 1993. Fearing that the gnatcatcher's listing would halt development, the state of California initiated its NCCP policy, incentivizing landowners to protect coastal sage scrub and create habitat preserves collated from public and private lands. The policy's ultimate goal is to conserve listed species and prevent future listings.

At the time of the gnatcatcher's listing, the NCCP policy was hailed as an innovative approach to biodiversity conservation in areas with intense human activity. But after nearly thirty years of implementation, it is time to ascertain whether the policy has prevented further declines in California gnatcatcher populations and other coastal sage scrub species. If so, the policy is a shining example of regional-level, development-friendly conservation. If not, it is a cautionary tale about underestimating the importance of the Endangered Species Act.

I began my scientific career in the early 1990s as a biology student at Pomona College, working with Professor William "Bill" Wirtz to investigate the California gnatcatcher's response to fire. As a requirement for my major in public policy, I interned at the Riverside County Planning Department, where I digitized maps of coastal sage scrub lost to development. I also attended planning meetings for the Western Riverside County Multiple Species Habitat Conservation Plan, one of the earliest plans approved under the NCCP policy. My research on the gnatcatcher and the NCCP policy became the subject of my thesis.

After graduation, I left Southern California for graduate school and a career, but I often thought about that little gray bird and the hopeful policy meant to protect it. Finally eligible for a sabbatical as a professor at Michigan Technological University, I returned to Southern California to see how the gnatcatcher and the policy had fared and whether the lessons learned in Southern California could be fruitfully applied to other rapidly urbanizing areas.

Acknowledgments

My first debt of gratitude belongs to Bill Wirtz, professor emeritus at Pomona College. Bill gave me my first research experience studying the gnatcatcher, advised my thesis, and became my friend. Bill and his wife, Helen, have launched many careers of Pomona College alumni, and the field of ecology is better off for it.

My gratitude extends to several other Pomona College professors, including Rick Worthington, who gave me a solid foundation in public policy, and Rachel Levin, who continues to provide guidance on a wide range of personal and professional matters.

Michigan Technological University and Pomona College provided invaluable sabbatical support. I had the pleasure to teach two classes of engaging and thoughtful Claremont Colleges undergrads who reminded me why I miss the Five Colleges campus so much. I am also grateful for the research librarians at Michigan Tech who helped me find hard-to-locate materials.

Friends and colleagues who provided feedback on specific chapters include Amy Louise Dyble, Peggy Flynn, Jennifer Lind-Riehl, and Char Miller. Laura Dyble, Jeanne Mayer, Christopher Lepczyk, and Michael A. Patten read the entire book and provided an abundance of insightful feedback. Sandrine Biziaux-Scherson provided her fantastic California gnatcatcher photos, Michael Arjun Banerjee assisted by "bluebooking" legal citations, and Jessica Alger created the wonderful maps in this book.

I owe a debt to all the bird watchers who participated in North American Breeding Bird Surveys and Christmas Bird Counts and entered their observations into eBird. Citizen scientists generally, and birders specifically, have become indispensable to environmental conservation, and I thank all of you for your dedication and generosity.

Jean Thomson Black, senior executive editor at Yale University Press, took a chance on a new author and provided limitless enthusiasm and good advice. Copyeditor Laura Jones Dooley and the Yale University Press staff shepherded this book toward publication.

Last, friends and family who helped my writing process in many ways: the Henquinet family; Jacqui (Canfield) Bolln; my sister, Sondra, for the "life" in the work-life sabbatical balance; my parents; and my son, Luukas, for his patience and keen appetite for adventure.

Abbreviations

CITES	Convention on International Trade in Endangered Species of Wild Fauna and Flora of 1975
ESA	Endangered Species Act
HANS	Habitat Evaluation and Acquisition Negotiation Strategy
HCP	Habitat Conservation Plan
IUCN	International Union for Conservation of Nature
MHCP	Multiple Habitat Conservation Program
MSCP	Multiple Species Conservation Program
MSHCP	Multiple Species Habitat Conservation Plan
NCCP	Natural Community Conservation Planning
NMFS	National Marine Fisheries Service
NRDC	Natural Resources Defense Council
Prop 13	California Proposition 13, passed in 1978
RCA	Western Riverside County Regional Conservation Authority
SANDAG	San Diego Association of Governments
SCAG	Southern California Association of Governments
USFWS	U.S. Fish and Wildlife Service

1

Setting the Scene

Southern California has long been an extraordinarily diverse place. About one hundred thousand years ago, in the Pleistocene epoch, when glaciers covered half of North America and California, the region was colder and wetter than today and was dominated by giants. Elephant-sized mammoths and mastodons browsed in redwood forests and oak woodlands, while camels and llamas grazed in clearings on brush and grass. These landscapes also supported 1,800-kilogram long-horned bison, which were 2.5 meters high at the shoulder and sported a pair of massive horns. Weighing 1,000 kilos, Jefferson's ground sloths roamed the area as Shasta ground sloths munched on Joshua tree fruit, spreading their seeds far and wide throughout the drier areas of the region.[1]

These giants were hunted by other giants. Saber-toothed tigers weighing 270 kilos and packs of 70-kilo dire wolves hunted prey in the forests and woodlands. The American cheetah chased down pronghorn across the grassy valleys, joined by 450-kilo American lions, considerably larger than the lions we see in Africa today. The massive short-faced bear also roamed the valleys and killed anything its 900-kilo, 3-meter frame could overpower. Mammoths and mastodons became trapped in tar pits, including those at La Brea Tar Pits and Museum in Los Angeles, attracting predators like tigers and dire wolves that sensed an easy meal (fig. 1.1). Occasionally these predators became stuck themselves and were pulled down to their deaths along with their prey. Condors fed on the dead before the tar took them, as did the Teratornis, a bird of prey with a 3.6-meter wingspan.[2]

About fifteen thousand years ago, the climate began to change. Worldwide, northern glaciers receded as the planet warmed by 1°–2°C over several thousand years—a blink of an eye in geologic terms but slow compared with today's pace of climate change. Southern California became hotter and drier yet still received the bulk of its yearly rainfall in winter. The pine forests abandoned the increasingly hot and dry valleys for higher and cooler elevations, ceding their place to oak woodlands, grasslands, and coastal sage scrub. Forest-dependent megafauna, such as the mammoths and mastodons, followed their habitat and comfortable temperatures to higher elevations, and many of their predators followed them.[3]

These giants also began to go extinct. One explanation suggests that declines in megafauna populations resulted from the loss of their habitats due to climate change. But another explanation, called the overkill hypothesis, lays the blame squarely on the spears of human hunters. During a volatile climate period called the Younger Dryas, receding glaciers in North America revealed an ice-free passageway across the Beringia land bridge between Siberia and Alaska down into North America. Humans streamed southward along this coastal passageway, reaching the American Southwest between 15,000 and 12,000 years ago. Most of the mammalian megafauna disappeared from California by 12,900 years before the present.[4]

In downtown Los Angeles at the La Brea Tar Pits and Museum, archaeologists have found the remains of a young woman who appears to have died from a blow to the head about nine thousand years ago. At that time, her people had advanced hunting technologies including equipment and strategy, honed for thousands of years on the Beringia land bridge. At the end of the Pleistocene epoch, human hunting thus either assisted climate change in the mass extinction

Figure 1.1. The author's descendent watched helplessly as a plastic descendent of a mastodon met its cruel fate in Rancho La Brea, downtown Los Angeles.
(Photo by the author, December 2016.)

of the megafauna or was the chief driver of it. One extinction can be directly laid at the feet of Southern California's early cultures: Law's diving-goose (*Chendytes lawi*) was hunted into extinction about 2,400 years ago.[5]

Two million years before the arrival of humans in North America, long before *Homo sapiens* existed, the California gnatcatcher (*Polioptila californica*) diverged from the black-tailed gnatcatcher (*Polioptila melanura*) in the interior deserts of the American Southwest and Mexico. During the same climatic event fifteen thousand years ago that brought humans down into Southern California, the California gnatcatcher made its way into Southern California and settled into the new, shrubby plant communities developing in semiarid spots along the coast. Having occurred sporadically around Southern California during the Pleistocene epoch, California sagebrush (*Artemisia californica*) became associated with sage (*Salvia* sp.), buckwheat (*Eriogonum* sp.), and other aromatic shrubs at about this time, forming an identifiable habitat type—coastal sage scrub. Humans began to manage this shrubby, flammable habitat to favor game populations and edible plants through such practices as purpose-set fires, tilling, and irrigation. The California gnatcatcher settled into the coastal sage scrub and began its fifteen-thousand-year relationship with humans in a landscape modified by human activities.[6]

10,000 Years before Present to 1800s CE

About ten thousand years ago, the global climate began to stabilize. Around this time, dense congregations of humans were forming in the Middle East, where they domesticated plant species that would blossom into agriculture. In coastal Southern California, hunter-gatherers diversified into tribes with unique languages: the Chumash, Tongva (or Gabrielino-Tongva), Tataviam, Cahuilla (or Iviyuqaletem), Luiseño, and Tipai-Ipai (or Kumeyaay). They lived in seasonally nomadic, small groups, spread across the region into which California gnatcatchers had expanded. Tribes processed acorns and pine nuts, practiced advanced horticultural techniques, harvested deer, fish, and quail, and used asphalt from the tar pits as glue and waterproofing. Their use of fire to keep the grasslands open and promote game species, such as antelope, mule deer, and rabbits, maintained a diverse patchwork of habitats throughout the region. Sometime between 700 and 1000 CE, humans in Southern California began to experience scarcity of their natural resources due to overharvesting, severe droughts linked to the global Medieval Climatic Anomaly, or both. Evidence of this period of declining health and violent conflicts among humans is clearly observed in the bones of their remains.[7]

The Southern California region has remained relatively dry since the global climate stabilized, with the exception of a 150-year stretch of wet weather about 8,200 years ago. The chaparral and coastal sage scrub plant communities flourished in the warmer, drier climate, diversifying into habitat varieties based on soil type, microclimate, and proximity to the coast. During this period of stabilized climate, as patches of coastal sage scrub took root among the grasslands in the valleys and the chaparral in the foothills, the California gnatcatcher expanded its distribution northward up to what is now Ventura County in the United States, and down to the southern tip of the Baja California Peninsula in Mexico. Stretched along this long, narrow range, California gnatcatchers began to differentiate due to their adaptations to local environments. Whether this geographic differentiation has produced distinct varieties that can be listed under the Endangered Species Act (ESA) is a key question explored in this book.[8]

The arrival of Europeans began the second biodiversity and first cultural catastrophe in the region. The first European in the area now known as Los Angeles was Juan Rodríguez Cabrillo from Spain, who explored the area in 1542–1543. Cabrillo first encountered the Chumash and Tongva, coastal tribes of about ten thousand people at the time. Cabrillo and other Spanish explorers called the region "California" from a Spanish novel about a beautiful island of the same name, ruled by the pagan Queen Califía and her army of black female warriors (yes, the archetype of DC Comics' *Wonder Woman*). Cabrillo named what is now San Pedro Bay (near downtown Los Angeles) as La Bahia de Los Fumos, or the Bay of Smokes, referencing the layer of haze created by thousands of indigenous fires trapped in the bowl-like geography of the area. The name was prescient of the heavy smog that would smother the region in the mid-1900s.[9]

During that time of early European encounters, the Tongva were seminomadic hunter-gatherers living across over 10,000 square kilometers throughout present-day Los Angeles and Orange Counties. Their population was roughly five thousand people, living in scattered villages of fifty to five hundred. Farther inland, at the eastern limit of the California gnatcatcher's range, the Cahuilla lived across 6,000 square kilometers between the San Bernardino Mountains and Borrego Springs. To the south, the Tipai-Ipai's territory encompassed many of the spots where California gnatcatchers are reliably spotted today, such as at Torrey Pines State Natural Reserve.[10]

During the 1700s, conflicts between Spanish colonists and indigenous tribes increased with the number of established Spanish missions and cattle ranches. The Missions period was an era of catastrophic decimation of the indigenous population. Tribes in the region dropped from roughly 30,000 people in the 1760s to 1,250 by 1910, which was cataloged by the gruesome ratio of baptisms to burials conducted by missionaries. Missions resettled indigenous people into work camps with inadequate food and sanitation, driving death rates upward due to malnutrition and disease. Inland tribes fared better than coastal tribes, as missions were clustered on the coasts in areas with the highest density of indigenous people and water resources. The Tipai-Ipai fared better than most tribes due to their resistance to missionaries and missionary work.[11]

Owned by the missions and left unfettered, livestock herds damaged natural resources needed by indigenous people still outside of the missions' reach. Multiple years of good weather encouraged missions to overstock cattle, leading to severe habitat degradation during drought years. Exotic grass seeds were brought in accidentally with cattle feed and became invasive, including velvet grass (*Holcus lanatus*) and wild oats (*Avena* sp.). Contemporaneously, Spanish pueblos produced large amounts of beans, corn, and wheat in the fields surrounding the indigenous communities, further restricting their access to natural resources. Although only a few thousand white settlers lived in the region during this era, their land-use activities had a massive impact on Southern California landscapes and tribes.[12]

In 1821, when the region officially became part of the new, independent country of Mexico, the land holdings of the Spanish missions were relinquished to private landowners through a usufruct system (literally, "use of the fruits"). The government transferred land ownership to private owners under the condition that the land was used for production, incentivizing the conversion of natural habitats to agriculture, grazing, and other economically productive uses. The Mexican government consolidated landholdings into private *rancheros* distributed among a small number of owners; fewer than eight hundred people owned over 3.2 million hectares. These massive rancheros were the ancestral roots of the modern development industry's large real-estate holdings in Southern California, particularly in Orange County.[13]

Landowners continued to expand their cattle herds into the valleys and foothills. Indigenous people, having lost numbers and power, were subjected to unrelenting violence and relegated to indentured servitude as farmhands on the new private ranches. Thirty years later, the region fell into the gravity well of the quickly expanding United States; California became a state in 1850. Arriving from the east, U.S. law quickly stripped the indigenous people of all rights and protections. Between the U.S. Census surveys in 1860 and 1870, the indigenous populations in San Bernardino and San Diego Counties dropped from about 6,100 to 28.[14]

The 1851 California Land Act transformed the Spanish-Mexican usufruct system to the American fee-simple system (private property), establishing a property system that would transform the region once again. With the first successful orchard in Riverside in 1873, orange groves quickly replaced ranches. Orchard owners aimed to "redeem sage and sand and glorify it" with orange trees from horizon to horizon. Alfalfa and beets took over the valley floors, and citrus orchards replaced chaparral and sage scrub at the base of the foothills as "a dark-green horseshoe curve around the rim of the valleys." The completion of two transcontinental railroads (in 1876 and 1886) ushered in a century of rapid human population growth and the conversion of native habitats and farm fields to urban land uses. Shipments of sun-drenched oranges and lemons supplied to the Chicago World's Fair and East Coast markets beckoned a wave of migrants westward. The California gnatcatcher and other native species began to feel the squeeze.[15]

1900s to the Present

Popular wisdom advocates that the infamous Los Angeles sprawl was set into motion by freeway expansion and a car-obsessed population boom after World War II. In fact, the region's sprawling suburban landscape was initiated several decades earlier, with an interurban streetcar system and an influx of easterners with a penchant for single-family homes. Each period of sprawl and population growth was fueled by the desire for a suburban lifestyle, encouraged by land developers who converted open space into cash, and enabled by regional plans that prioritized suburbs over all else.[16]

The first episode of rapid sprawl began between 1880 and 1910 with an explosion of low-density, residential subdivisions. The streetcar rail system allowed real-estate speculators to cater to middle-class transplants from eastern cities who wanted homes far from dirty, crowded commercial and industrial areas. While the streetcars usually operated at a loss, land speculation along the streetcar railways was hugely profitable, even to areas far from urban centers. By 1910, Los Angeles had the largest interurban streetcar rail system in the nation. By 1925, residents could use the system to commute from the San

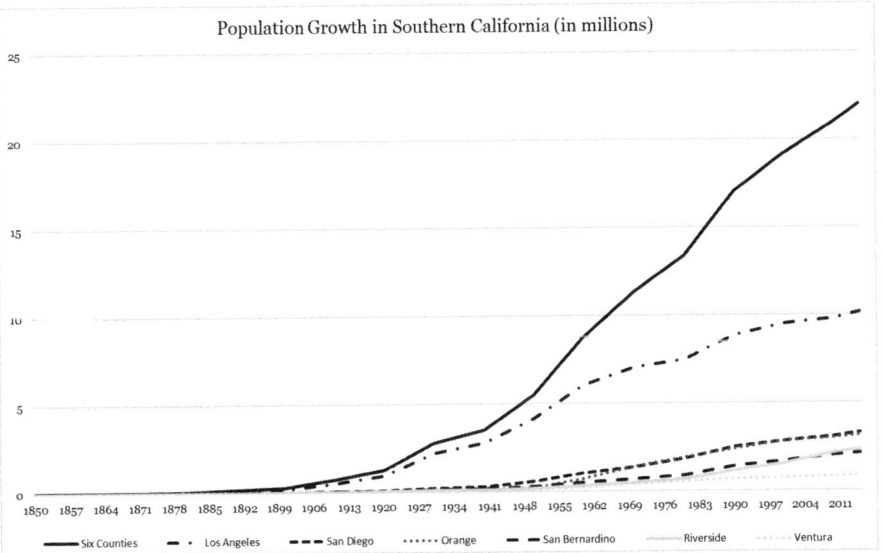

Figure 1.2. Human population growth in the six-county Southern California region, from 1850 through 2015.
(Data from U.S. Census Bureau.)

Fernando Valley to Orange County and from Santa Monica to San Bernardino.[17]

The second boom in population and development hit in the 1920s (fig. 1.2). Industries such as entertainment and oil and gas exploration began to decentralize, forming small urban satellites away from the central business districts. During this expansion, personal automobiles facilitated longer commutes to work. The upsurge of private cars was accommodated by urban planners who prioritized wide streets and plentiful parking in city centers and by developers who built single-family homes with driveways and garages. The nation's first regional planning commission began in Los Angeles during this time, dominated by growth-driven land barons: bankers, real-estate agents, and developers. By 1930, over 90 percent of the region's residential units were single-family homes (compared to roughly 50 percent in eastern cities), and half of all visits to downtown Los Angeles were made by car. Highways became less of a complement and more of a competitor to the interurban streetcar rail system, which was rapidly losing popularity among residents due to increasing fares and slower speeds.[18]

The decades between 1930 and 1950 saw a spike in the amount and size of publicly owned land designated as open space for recreation and conservation. During this time Death Valley and Joshua Tree became national monuments, joining four large national forests (the Angeles, San Bernardino, Los Padres, and Cleveland National Forests) that had been set aside in the late 1800s. But protected areas in the mountains and deserts did not protect the California gnatcatcher's coastal valley habitat.[19]

Fed by continuing immigration and migration from the east, the regional population continued to grow. Between 1900 and 1940, the population of Southern California grew by 1,142 percent (see fig. 1.2). In 1880, the city of Los Angeles had barely been a village, with a population of about twelve thousand. But between 1900 and 1940, the city grew by 1,536 percent, swollen by Dust Bowl refugees, outpacing regional growth over the same period. The combination of irrigation and refrigeration drove an expansion of orange groves into every fertile corner of Southern California. This explosion in orchards eliminated much of the California gnatcatcher's remaining habitat, which was likely why the species was common in the late 1800s and rare by the 1940s.[20]

The end of World War II and the booming economy of the 1950s sent millions of young families in search of housing. The ubiquity of personal automobile ownership became a priority in land-use planning and drove residential sprawl into the orange groves and remaining sage scrub. Ironically, advocates of suburbanization pushed car culture as a mechanism to provide more residents better access to nature. Attracted by the promise of mild weather, beautiful beaches, and economic opportunity, additional waves of migrants generated another doubling of the population in the region, from 3.5 million to 8.8 million people between 1940 and 1960. The value of houses far outpaced the value of orchards; by 1995, just 405 hectares of orange groves remained in Orange County. In Claremont, all that remains of the once-prominent citrus industry is the Packing House—once a factory floor where women carefully packed orange crates for distant eastern markets, now converted into trendy stores and restaurants. Rapid urbanization eliminated almost all of the coastal sage scrub that had remained after the citrus boom several decades earlier.[21]

Southern California's land barons would finally experience pushback during the environmental movement of the 1960s and 1970s, primarily from homeowners and environmentalists who were exasperated by the loss of open space (fig. 1.3). Throughout the state, the 1970s were marked by a series of citizen initiatives on local and state ballots to restrain growth and promote open space. By the late 1980s, this pushback became significant. Developers began to lose their grip on local politicians, who felt intensifying pressure from citizen ballot initiatives to place zoning restrictions on development. Even in Riverside County, where the local economy was almost completely dependent on the construction industry, a citizen's growth control coalition put on notice politicians who were cowing to the development industry. Ultimately, these coalitions fell apart under the relentless demand for housing.[22]

The lack of effective regional planning and environmental preservation has been obvious to many for a long time. As William Fulton artfully described the situation in *The Reluctant Metropolis* in 2001:

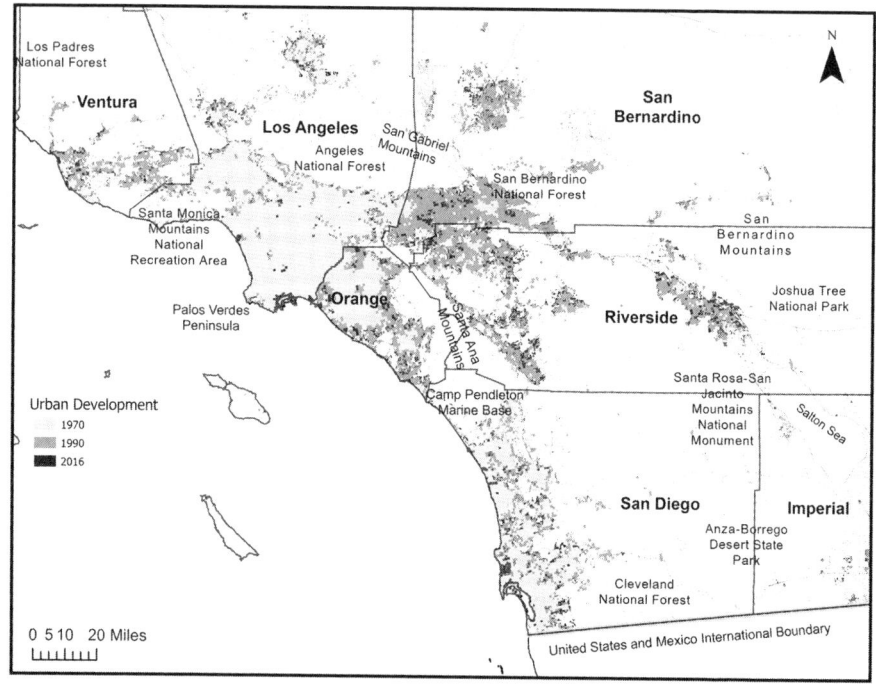

*Figure 1.3. After 1970, most urbanization occurred outside of the
major cities of Los Angeles and San Diego, sprawling into the
inland valley of Riverside and San Bernardino Counties.*
(Sources: National Land Cover Database, 2011; Historical Land-Cover Change and Land-
Use Conversions Global Dataset [1970, 1990]; UI-UC/ATMO, Department of Atmospheric
Sciences, University of Illinois at Urbana-Champaign. Map prepared by Jessica Alger.)

Especially in the go-go years of the 1980s, the growth patterns of the region were determined by an endless series of
local battles over pretty plateaus and dramatic vistas, in which
developers sought to gain enough political power to bull-
doze the landscape and environmentalists used every means
at their disposal to try to stop them. Inevitably, most such
disputes were settled by giving half a loaf to each side. The developer would scale down the project and dedicate, or sell, the

rest of it to some public agency for parkland. Which property was preserved and which was developed depended on the political dynamics of any given situation. In general, however, the resulting patterns of land development did not represent anybody's idea of "good planning."[23]

Even as far back as 1937, when California mandated that cities and counties adopt general plans, state politicians realized that some mechanism for regional-scale coordination was necessary for a variety of needs, including transportation, open-space provisioning, and water management, and also for effective, large-scale conservation policy. A state planning office was created in 1959 to oversee planning decisions as prescribed in these general plans. Despite this foresight, the office—now called the Office of Planning and Research—has never had sufficient power to sanction anything other than planned housing developments.[24]

Formed in 1965, the Southern California Association of Governments (SCAG) is a regional planning body that is ostensibly meant to coordinate development, open space, and other regional-scale issues. In truth, SCAG was not born from a burning desire to increase planning and coordination across the region. Rather, the association was created to qualify for federal funding for regional-scale transportation projects. As a governing board of more than eighty members, SCAG represents 184 cities, parts of six counties, and more than a thousand special districts and other quasi governmental organizations such as the Metropolitan Water District of Southern California. Most of the California gnatcatcher's range in the United States is encompassed by SCAG's oversight, plus parts of Riverside County represented by the Western Riverside Council of Governments, and San Diego County represented by the San Diego Association of Governments (SANDAG). In theory, these regional governmental associations are exactly what Southern California needs to manage population growth, urban sprawl, open space, and biodiversity conservation. In practice, these associations are often not taken seriously—even by their own members—and do little to generate open space or preserve nature. For example, in 1994, SCAG adopted a regional comprehensive plan that was divided into required core chapters and optional ancillary chapters: open space and other environmental planning concerns were in the ancillary chapters.[25]

Another source of fuel for the regional growth machine is Proposition 13 (Prop 13). Passed by tax-weary voters in 1978, Prop 13 constrained property tax rates on existing development across the state, strongly incentivizing municipalities to permit new development—and, hence, new sources of revenue—wherever and whenever they could. Combined with other revenue-limiting propositions passed in 1979 (Proposition 4), 1996 (Propositions 62, 218, and 268), and 2010 (Proposition 26), these propositions created a conservation paradox: to generate revenue for land acquisition for open space and habitat conservation, cities and counties have to permit new development.[26]

By the 1980s and 1990s, when endangered species' listings began to present serious obstacles to new development and, thus, new tax revenue, citizen groups supported the use of bonds for open space and habitat conservation in order to circumvent the restrictions of Prop 13. Environmental organizations began land acquisition campaigns funded by donations to do the same. At the same time, local governments were greenlighting larger and more sprawling development projects, particularly big-box retail in city peripheries, which brought in higher retail tax revenues. When a recession in the early 1990s erased half a million jobs, the tension between economic prosperity provided by growth and development and environmental protection through open-space preservation became palpable. William Fulton aptly described Southern California as "a churning bundle of consequences, a 'growth machine' throwing off side effects left and right." Other authors have characterized the past few decades of land-use planning as a period of "fiscalization of land use," in which land-use decisions are driven overwhelmingly by the desire to increase a locality's tax base above all other priorities. Much of the headwinds that environmentalists, homeowners, and planners face regarding open-space and habitat conservation emanate from the giant tax-base vacuum that Prop 13 created. These headwinds also affect the Natural Community Conservation Planning (NCCP) process, as I explain in chapter 6.[27]

Apart from the mountain ranges and the ocean, the Southern California of today would be almost unrecognizable to the ancestors of the Tongva, Tipai-Ipai, and Chumash who still occupy the region, speak their languages, and fight the U.S. government for recognition and rights. The region would be barely recognizable to the Spanish missionaries. The landmark Gage Mansion, built in 1795 and the oldest house still standing in Los Angeles County, is now surrounded by trailer homes. The missionaries and the orange growers had little idea that the region would explode into the second most populated area in the United States, enduring exponential population growth for decades on end (see fig. 1.2). All of these new people need water, houses, factories, office parks, roads, and parking lots—lots of parking lots, in fact, as 14 percent of Los Angeles is dedicated to parking.[28]

At present, a third of Los Angeles County is dedicated open space, much of that in the Angeles National Forest in the mountains. Nearly half of San Diego County is open space—much of it in mountains and deserts or within Marine Corps Base Camp Pendleton. Up until 2002, Los Angeles County had the highest number of protected areas of any county in California and had the third-highest per capita protected area. Some of this open space was the result of the 1965 Quimby Act, which required developers to set aside open space in every development project. These disconnected, small parks are scattered throughout areas of high real-estate value, and thus, Los Angeles, San Diego, Orange, Riverside, and San Bernardino Counties have the most expensive protected areas in the state.[29]

Most of the land designated as open space exists on the periphery of urban areas, far from people. The current landscape thus reflects decades of loss of—and separation from—the natural beauty that attracted those millions of newcomers to Southern California in the first place. The scarcity of open space generates conflict among different uses, especially when open space is meant to double as protected habitat for at-risk species (box 1.1). As I recount in chapter 6, one small remnant of coastal sage scrub set aside for California gnat-catchers in San Diego County is being loved to death by mountain bikers. One parent explained, "I want my kids and my friends' kids to be able to go out and ride in dirt.... There's no more dirt left. It's all concrete."[30]

Although preserved open space is one positive consequence of the Natural Community Conservation Planning policy, the policy's main focus is to protect California's natural heritage for future generations, a considerable challenge. In the conterminous United States, California is home to the highest number of species threatened with extinction: 283 plant species are listed on either the state or federal ESA (or both), as are 154 animal species. San Diego County has more federally listed species than any other county in the continental United States. Thirty-two species in the state have gone extinct, mainly due to habitat loss. California's notable aquatic diversity has been hit particularly hard by hydrological alterations for human water use; in the deteriorating San Francisco Bay estuary, the endangered delta smelt (*Hypomesus transpacificus*) is now feared extinct in the wild. California's

Box 1.1. North Etiwanda Preserve, Rancho
Cucamonga, San Bernardino County, March 5, 2017

Even at dawn on a weekend, the park was crowded with walkers and joggers. My son and I were greeted by a good omen: a cartoon California gnatcatcher perched on a yucca plant, prominently displayed on the park's entrance sign. At a fork in the trailhead, my son demanded that we take the high trail, away from a great patch of coastal sage scrub and up into the foothills. I was sure that we were going in the opposite direction from any potential gnatcatchers. But we found a pair about halfway up in a brushy wash right next to an area that recently burned—California sagebrush was bursting through the soil after the winter rains. The gnatcatcher pair was 200 meters from the remains of an old cabin built by white settlers on what was once the Tongva village of Kuukamonga. I returned to this park a week later to walk the low trail that ran parallel to a newer residential neighborhood. I assumed that I would find another gnatcatcher pair or two in the nice-looking sage scrub along the trail. But the noise from the neighborhood, roads, and overhead high-voltage transmission lines made it difficult for me to hear any gnatcatchers even if they had been there.

reptiles compete for space with solar and wind-energy developments in deserts and high ridges. Among California's mammals, nineteen species and eighteen subspecies are listed under the federal ESA. The extinct California grizzly bear (*Ursus arctos californicus*), once numerous in Southern California, was an important initiator of soil disturbance as it dug for small burrowing mammals and tubers. California's flag is the only government banner in the world to enshrine an extinct species; the last California grizzly bear was seen in 1924, years before the ESA could have saved it.[31]

California's astounding biodiversity and endemism (species found nowhere else) is a consequence of its climate. The state's cool, wet winters and hot, dry summers are typical of Mediterranean-type ecosystems, which represent less than 5 percent of the world's land area but support 20 percent of the world's plant species. Only five regions in the world fit this profile: the Mediterranean Sea basin (supporting 60 percent of the global total of this ecosystem type); southern Australia (22 percent); California (10 percent); central Chile (5 percent); and the Cape Province of South Africa (3 percent). All of these regions support extremely high biodiversity and endemism, classifying them as biodiversity hotspots. These hotspots are often prioritized when creating new protected areas and initiating research projects, because they represent both tremendous biological uniqueness and a high risk of loss. All Mediterranean-type ecosystem regions are experiencing biodiversity losses due to human land-use activities, such as urbanization, agriculture, and grazing, as well as pressure from invasive species and climate change. Unfortunately, species endangerment and extinction are not limited to California and other Mediterranean-type ecosystems. In 2019, scientific reports of large-scale declines in the abundance of birds and insects made headlines—one study estimated that North America has lost over a quarter of its birds since 1970, representing a loss of about three billion birds. Changes in land use and land cover are the primary driver of loss of biodiversity almost everywhere in the world, ushering in the sixth mass extinction on the planet. As a land use, urbanization drives more species to the precipice of extinction than almost any other human endeavor (agricultural practices cause more).[32]

One of Southern California's most poignant legacies is the sacrifice of its spectacular and abundant biodiversity to urbanization, driven by land-use policies that were either ambivalent about habitat destruction or actively encouraged it. By the 1990s, the region was desperate for a champion to preserve the remainder of Southern California's natural history. The listing of the California gnatcatcher under the ESA in 1993 made the gnatcatcher a serendipitous guardian of the region's biodiversity, with gnatcatcher populations ascribed to the success or failure of a novel state strategy, the NCCP policy. The California gnatcatcher quickly became a controversial conservation ambassador, expected to shield the region's biodiversity from the bulldozer's blade under its tiny wings. After thirty years, it is time to ask whether this approach has worked to stem the loss of regional biodiversity or whether more is needed to preserve biodiversity in a region that continues to urbanize. The lessons learned from California's conservation policies are applicable to many other rapidly urbanizing areas around the world.

2

Essence of a California Gnatcatcher

To identify the California gnatcatcher's conservation needs, we must first understand the traits of the gnatcatcher's life history. These traits include their breeding behavior (how quickly they can increase their numbers and how much nesting territory they need), their niche and foraging behaviors (what they need to survive), and their distribution (where they have been found in the past and where they are found now). These characteristics influence how easy it can be to push the species toward extinction or recovery. Species that produce few offspring and require a long time to raise them are unlikely to maintain their numbers under chronic threat or rebound quickly in good conditions. Species that don't disperse well are unlikely to find new habitat patches quickly. Luckily, gnatcatchers can both build up their numbers quickly and disperse well enough to find new patches of habitat. Although gnatcatchers are a relatively short-lived species, in years with good weather and in high-quality habitat, California gnatcatchers have the capability to increase their population rapidly. During less favorable conditions, they maintain their numbers or decline.[1]

Adult gnatcatchers require about 4 to 8 hectares of coastal sage scrub for a breeding territory, but smaller habitat patches are valuable to immature gnatcatchers as stepping stones to find their own breeding patches. Juveniles possess the capacity to travel many kilometers through or around urbanized areas to find new habitat, and an urban landscape can support gnatcatcher dispersal either with coastal sage scrub and riparian corridors or sufficient urban vegetation and green space. Conversely, the capacity of gnatcatchers to produce many new gnatcatchers is hampered by nest predators and, to some extent, the brown-headed cowbird (a brood parasite), suggesting that smart urban planning is necessary to separate breeding patches of high-quality coastal sage scrub from human activities that increase the population of nest predators. Luckily, some human activities are compatible with gnatcatcher conservation; for example, dog-walking trails can reduce nest predation in California gnatcatchers.[2]

But before we discuss the star of this show in depth, a brief review of some key terminology is in order. The term *eggs* is self-explanatory. A *clutch* is a batch of eggs that are all laid, incubated, and hatched at the same time. A female will typically lay one egg per day until she determines that the clutch is complete, and then she and the male will incubate the eggs. The words *hatchlings* and *nestlings* are generally interchangeable, although some use the word *hatchling* for chicks that are up to just a few days old, reserving *nestling* for older chicks with feathers. *Fledglings* are chicks that have left the nest but are still hanging around, expecting to be fed by their parents. *Young of the year* are the juveniles that hatched in that year and have not yet grown their adult breeding plumage. *Adults* are birds that were hatched a year or more before. Generally, gnatcatchers must live through one winter before they are considered breeding adults, although juveniles from early spring clutches have been observed breeding in late summer of their hatching year.[3]

The Distinctive Sound of a Gnatcatcher

The California gnatcatcher has a distinctive voice, which is extremely helpful when we conduct field surveys to estimate its population and to identify patches of habitat where gnatcatchers are present. Adjectives to describe the California gnatcatcher's songs and calls are all variations on a theme, from my interpretation as "whiny" to Robert Woods's description in the 1920s as "querulous, thin, and plaintive":[4]

> It was at about this time [late August, when the male's plumage was changing] that I noticed the only indication of a song that I have ever detected on the part of this species— possibly only a musical inflection of the usual call. Practically all the utterances of the adults seem to be variations of the rather prolonged, mewing call note, which is often repeated from two to four times in succession. The call of the female is thin and plaintive; that of the male is usually stronger and heavier, and sometimes scolding, but is nearly always characterized by the complaining tone which distinguishes the voice of this species from that of the Western gnatcatcher (*Polioptila caerulea obscura*).[5]

All species of gnatcatchers have a "mewing" quality to their song, but the California gnatcatcher's is the most distinctive, at least to my ears.[6]

Aside from their distinctive song, California gnatcatchers have a variety of calls, including a territorial song, a warning call, and a summons to wayward fledglings. Whenever I hear that "summons to youngsters," I always pause for a few moments to count the fledglings as they appear. Given their songs' lack of appeal, you will not see gnatcatchers singing away in a pet store any time soon, which is another piece of good news for their conservation. Similarly, there isn't much meat on a bird that is 12 centimeters long and weighs 5–6 grams. Overharvesting for the pet trade and for food is a major cause of species endangerment and extinction across the world, but not for the California gnatcatcher.[7]

Breeding and Life Cycle

When determining the population size and reproduction rate of a species, one challenge is the ability to count the number of juveniles that survive to the next breeding season. One might ask how ornithologists can distinguish males from females and from juveniles (which look like females), especially when they all look similar in the nonbreeding season. One way is by feather condition. An individual's age and gender can be determined based on the condition of its plumage and by the time of year. Most bird species molt—lose and replace their feathers—twice a year. In gnatcatchers and many other birds, the first molt happens quickly before the breeding season so that individuals can look appealing to mates, but also, for migratory birds traveling long distances, to provide fresh feathers. The second molt happens gradually once breeding is done, which benefits migrants for their flight to their nonbreeding grounds and residents for warmth during the winter months. Juveniles will not go through a fall molt. Instead, they keep their first feathers until the start of the next breeding season, so juveniles look increasingly ragged as fall and winter progress. Thus, when you're standing on a small, barely mentionable patch of coastal sage scrub in winter and you see a raggedy-looking gnatcatcher, you can be fairly sure that it's a juvenile dispersing away from its parents, looking for a territory of its own.[8]

Male gnatcatchers grow their distinctive black cap for the breeding season in the late winter molt (fig. 2.1). They slowly lose the black cap to all gray feathers in the fall, although they keep a black streak above their eyes, helping to distinguish them from females in the nonbreeding season. Both sexes have a marked white edging along the outer tail feathers, which gradually wears off as the feathers age. All California gnatcatchers keep a distinct white eye-ring year-round. These cap, eye, and tail markings are used to distinguish males from females, juveniles from adults, and California gnatcatchers from other gnatcatcher species. Not only do these differences in feather markings and condition permit us to identify sex and age, but they also allow us to determine how many breeding pairs can fit into a patch of habitat. We can also identify those extra birds in a habitat patch as fledglings or young birds moving across the landscape to find new habitat.[9]

Fecundity and Nesting

To determine the population dynamics of a species, we measure the species' *fecundity,* defined as how many offspring it can produce and how quickly. Species that produce a lot of young can rebound quickly after dire conditions and are usually easier to manage and conserve because they can replace themselves quickly. Species that can reproduce within a year or two of hatching can also increase their populations quickly relative to those species that require several years before individuals mature to breeding age. That said, humans excel at reducing even the most abundant and fecund species to zero. For example, the extinction of the passenger pigeon (*Ectopistes migratorius*) a century ago is a testament to our unique and unfortunate ability.[10]

The breeding season of the California gnatcatcher is roughly five months long, from early March through early August. As a consequence, gnatcatchers raise their young across highly variable weather conditions—from cool, moist springs to hot, dry summers. Robert Woods was the first to document the breeding success and nesting habits of adults as well as the behavior of juveniles in the period before they become adults. During the 1920s he followed a pair of gnatcatchers around their breeding territory near the town of Azusa, California,

Figure 2.1. California gnatcatcher male (above) and female (below) carrying a caterpillar in her beak.
(Photos: Sandrine Biziaux-Scherson.)

for several months at a time. He determined that gnatcatchers can lay up to three clutches per year, with two to four eggs per clutch. Since his work, gnatcatchers have been observed trying to build and fill a nest up to ten times in one season. Gnatcatchers will start a new nest and clutch if the first nest was poorly constructed, too conspicuous, or damaged by heavy rainfall (more likely early in the breeding season) or if the eggs or nestlings were killed. Gnatcatchers with more breeding experience will lay smaller clutches during extremely wet weather periods to reduce incubation failure.[11]

Gnatcatchers lay two to six eggs per clutch, depending on weather conditions and other factors. On average, a female lays a total of about nine eggs across all nests in a season, with a maximum observed of fifteen eggs. About 90 percent of eggs hatch, unless they are improperly incubated or eaten by predators. The probability that an egg survives to become a fledgling declines with the number of eggs per clutch, so there is a trade-off between the number of eggs and the survivorship of the young. Averaging across all of the eggs, clutches, and hatchlings produced over a season, each pair of gnatcatchers produces about three fledglings per year. In years with good weather and ample habitat, gnatcatchers may produce more than this, and in bad years, less than this. Three fledglings per pair is enough to replace the male and female, plus one extra. If a lot of those extra fledglings survive to join the breeding population in the next year, gnatcatcher populations increase. Along with fecundity, mortality rates govern population trends—in one study, annual mortality was as high as 60 percent for adults and 90 percent for juveniles. On average, adult gnatcatchers live about two to three years, and eight years is the likely upper limit.[12]

Finding gnatcatcher nests is no small feat. Gnatcatcher nests are tiny, and the pair sticks their tiny nest right in the thickest part of the thickest bush they can find. Given time and patience, you can find a nest when you see adult birds flying into the same bush with nesting material or insects in their beaks—dead giveaways for a nest under construction or hungry young. California gnatcatchers build their nests out of grass, thin bark fibers, and other plant materials. They often line the inside of the nest with feathers, fur, flower petals, moss, or other soft debris. The inside of a nest is only about 4 centimeters wide and not quite 3 centimeters deep. Yet when a gnatcatcher is sitting on its eggs, you can still see only the top of its head and tail. California gnatcatchers prefer to nest in the middle of California sagebrush, California buckwheat (*Eriogonum fasciculatum*), or chamise (*Adenostoma fasciculatum*) bushes, which grow about waist-high and thick. In a pinch, or if they are inexperienced, gnatcatchers will build nests in almost any shrub, including laurel sumac (*Malosma laurina*), Brewer's saltbush (*Atriplex lentiformis* ssp. *breweri*), and even Australian bottlebrush (*Melaleuca nesophila*). Gnatcatchers will nest near areas with considerable human activity; nests have been found in bushes next to trails, roads, construction sites, and airports.[13]

Parental Care

Gnatcatchers are egalitarian. Both parents incubate, although males spend less time on or near nests, particularly toward the end of the breeding season. Both parents feed the young before and after fledging, and both chase off other birds that might usurp their territory or animals that might kill their young. Once incubation starts, eggs usually hatch after fourteen days. Nestlings remain in the nest for another two weeks and then fledge. Fledglings hang around the nest to be fed for another few days up to a week and then begin to explore the territory. Parents feed fledglings usually for about four weeks, until the parents get tired of feeding them and chase them off—often because they are busy with the next clutch. And there's no hanging out in the basement; parents aggressively chase away juveniles once they become adult sized. Juveniles from the same clutch may sometimes remain in a flock and investigate adjacent areas together for suitable habitat, although they then disperse alone to find their own patches of habitat as they mature and prepare for the next breeding season.[14]

California gnatcatchers tend to stay put in an area once they've reached adulthood and defend territory year-round. On average, a breeding pair of gnatcatchers needs about 4 to 8 hectares of coastal sage scrub for a territory. In highly productive coastal areas, territories as small as 1 hectare have been documented, whereas in poor inland habitat, one territory might encompass up to 18 hectares. Gnatcatchers expand their territory boundaries when they have fledglings needing to be fed but relax their boundary defense during the nonbreeding season. Overall, these territories are quite large for such a small bird and, thus, require female gnatcatchers to help males defend them. The size of these territories and the effort needed to defend them suggest that the resources provided in coastal sage scrub are not overly abundant for even small birds. Lack of abundant resources is perhaps why gnatcatchers so readily feed in adjacent habitats, especially riparian zones that support more insects.[15]

Dispersal

In 1928, Robert Woods described the homebody nature of the California gnatcatcher as one of its most defining characteristics:

Probably no other California bird is so strictly confined to the brush-lands as is this. The Cactus Wren shares the same territory, but often comes about houses and dooryards, and sometimes builds nests in orange groves and around build-

ings, while the Black-Tailed [now California] Gnatcatcher almost invariably turns back when it reaches the limits of the
natural vegetation. The most striking contrast between this
species and the Western [now blue-gray] Gnatcatcher is that
of the restricted habitat of the former and the wide-ranging
habits and distribution of the latter.[16]

The California gnatcatcher is considered to be an umbrella, flagship, or surrogate species for the conservation of coastal sage scrub
due primarily to the gnatcatcher's affinity for it. Using the gnatcatcher
as a surrogate species assumes that protecting enough coastal sage
scrub for the gnatcatcher should also protect other species that are
dependent on coastal sage scrub habitat. I will address this assumption of surrogacy in chapter 7.[17]

Field ornithologists have long suspected that adult gnatcatchers
don't travel far once they have established a territory. By sitting quietly
with a pair of binoculars, we can often discern the boundaries of a
breeding pair's territory by the border disputes that erupt with the
adult gnatcatchers in adjacent territories. Year after year, unless a fire
or a bulldozer has destroyed the habitat, we can return to the same
patch and find a pair of gnatcatchers in roughly the same spot. Of
course, since all gnatcatchers look alike, we can't be sure that the male
and female building a nest in that territory are the same ones that were
there a year or even a day before.[18]

The practice of bird banding has resolved the difficulty of identifying individual birds. Banding—or ringing in British English—
refers to placing one or more bands on a bird's legs to provide each
bird with a unique identification mark. Banding is now common
throughout the world, allowing researchers to distinguish individuals and track their movements. Every bird that is banded is required
to have one steel band that has a unique number for the banded

individual of that species. Additionally, banders might place one or more colored bands on each leg, and each bird in the study receives a unique combination of colors. Unique combinations of colors for each bird help researchers identify and track specific individuals by sight rather than having to catch them again. The numbers from the steel band and the colored band combinations are reported to a central repository that manages the data, for example, a natural resource agency, which in the United States is the U.S. Fish and Wildlife Service. Bird banders record data for each banded bird (weight, sex, size, plumage color, breeding status, and location of banding), which are sent along with the band number to the central repository. Anyone who finds a banded bird can report the band numbers and colors to the repository and can learn where and when the bird was banded and last seen. This system allows researchers to determine how far their banded birds travel, what habitats they use, and how long they live. When nestlings are color-banded along with their parents, researchers can determine the relatedness of a population of birds and determine how far each juvenile travels to seek new habitat (called *natal dispersal*) once they leave their parents.

Once the parents kick the juvenile gnatcatchers out of their natal territory, the juveniles spend the late summer and fall bouncing among habitat patches, identifying vacant areas where there are no adults defending occupied territories. Until recently, researchers generally assumed that juvenile gnatcatchers did not disperse far from their nest site and would not move through habitat other than coastal sage scrub. California gnatcatchers are not strong, confident fliers. They flit and skulk between bushes, and when they do manage to make a flight over the top of them, they look as if they're at constant risk of stalling out. If gnatcatchers have poor dispersal capacity, they would require an extensive network of coastal sage scrub corridors to link habitat patches and prevent the isolation of small populations of gnatcatchers.[19]

In 1998, a trio of banding studies suggested that some juveniles could disperse long distances through habitats other than coastal sage scrub. Eric Bailey and Patrick Mock banded one hundred juveniles on the Palos Verdes Peninsula in western Los Angeles County, an isolated patch of coastal sage scrub sandwiched between the Pacific Ocean and a sea of Los Angeles sprawl. Among the banded juveniles they were able to find again were a few that had wandered up to 5 kilometers away from their natal territory, although most had stayed closer to home. J. Paul Galvin observed similar dispersal behavior in Orange County. He found one banded juvenile in a new habitat patch 7.5 kilometers away from its natal territory, but the vast majority of the relocated banded juveniles stayed within a kilometer of their natal territory. Kristine Preston and others conducted a banding study in San Diego County and recorded one banded male in a new breeding territory 4 kilometers away from its natal territory.[20]

Banding data have also revealed that juvenile gnatcatchers are not restricted to coastal sage scrub as they move through the landscape to find their own territories. At least nine of the hundred banded juveniles from the Palos Verdes Peninsula study settled in habitat that was separated from their natal territory by suburbs, urban parks, golf courses, and even an interstate highway. Tracing the most parsimonious route that each of the nine juveniles could have taken—that is, sticking to low-elevation, well-vegetated areas—the authors estimated dispersal distances as short as 500 meters and as far as 7 kilometers.[21]

One limitation of banding studies is the inability to track or find every banded individual, especially dispersing juveniles. We know how far they move only if we find them again, and given the vast area these ecologists are trying to cover, the odds of finding juveniles again are low. The rapidly advancing field of genetics can help with this logistical limitation. Amy Vandergast and colleagues used DNA collected in the blood and feathers from 268 individual gnatcatchers living around Southern California to identify their genetic first-order relationships—parents, offspring, and full siblings. The genetic data suggested that, while most gnatcatchers settle within 4–5 kilometers of their natal territory, a few move much farther—far longer distances than previously known. (This dispersal pattern is fairly common among terrestrial animals.) In fact, their data revealed evidence of a few individuals that were over 130 kilometers away from their family members. Gnatcatchers that made these long-distance journeys crossed through urban areas where at least 10 percent of the landscape consisted of patches of coastal sage scrub or similar habitats. These data don't necessarily mean that one gnatcatcher traveled 130 kilometers, since it is possible that two family members could have each trekked about 70 kilometers in opposite directions. Notably, long dispersal distances were more common where there was sufficient habitat and green space for gnatcatchers to use as stepping stones. These banding and genetics studies all emphasize the importance of designing networks of protected habitat that are well connected by small habitat patches for stepping stones and are surrounded by an abundance of vegetated open space in the urbanized landscape.[22]

I can add my own anecdote of juvenile California gnatcatcher dispersal and frustration over losing valuable connective patches to development. The Bernard Field Station at the north end of the Claremont Colleges campus sits between the vast urban expanse of the Pomona Valley and the San Gabriel foothills and supports 35 hectares of mixed coastal sage scrub, grasslands, and oak woodland. One day in 1994, while teaching an outdoor biology laboratory at the field station, I heard and saw a California gnatcatcher in a patch of coastal sage scrub. By that point, I had been studying California gnatcatchers every week for three years, so I knew these gnatcatchers well.

At that time, the faculty, students, and alumni were engaged in a bitter dispute with the Claremont Colleges Consortium over its plans to develop 15 percent of the Bernard Field Station for a graduate school focused on life sciences. This was one small example of the compulsion to destroy native habitat for more development that was playing out all across Southern California, one patch-to-building conversion at a time. The consortium viewed my observation of a federally listed species on the field station as a little too convenient, just one of many stall tactics of the "Save the Field Station" group. The California gnatcatcher I saw was likely a juvenile dispersing along the narrow strip of coastal sage scrub at the base of the San Gabriel Mountains to (I hope) larger, more productive patches for a breeding territory of its own.

Several years later—too late for my vindication, since I had moved to Knoxville, Tennessee, to start graduate school—Jon Atwood published a paper in 1998 reporting a California gnatcatcher sighting in north Claremont: "The species was documented for Los Angeles County north of Claremont in 1994 [by Jon Atwood] along the alluvial fan (Calif. Dept. of Fish & Game Natural Diversity Database). From north Claremont east through Etiwanda toward north Rialto, along the southern slope of the San Gabriel Mountains, extends approximately 30 kilometers of Riversidean alluvial-fan sage scrub, most of which is not regularly surveyed for California gnatcatchers."[23]

Was it the same juvenile bird I saw? It wasn't banded, so there's no way to know. But I'm sure the college consortium heard my "I told you so!" all the way from Knoxville.

I am happy to report that as of this writing, the Bernard Field Station still supports 35 hectares of habitat, having so far avoided devel-

Box 2.1. Claremont Hills Wilderness Park,
Los Angeles County, May 7, 2017

I arrived at the park just before dawn, thinking I would beat the crowds and have the place quiet and to myself. Not a chance. As I pulled into the parking lot, two clusters of joggers were setting off, and more cars were coming in behind me. I took the Cobal Canyon side of the 6.5-kilometer loop trail, since the eBird database noted that a California gnatcatcher was spotted here in 2013. Though this was a birding hotspot, there were no other records of California gnatcatchers. Quite soon I realized why; the park was dominated by large riparian zones in the canyons, and a mix of grassy oak woodlands and chaparral on the slopes and ridges. There was a bit of coastal sage scrub lower down among the oaks, but this quickly gave way uphill to its chaparral cousin. The park was a bit too high in elevation to support coastal sage scrub, although this might change as the region warms and dries out over the next few decades. The park was an awesome place to go birding due to the diversity of habitat types, but perhaps not an awesome place to raise a family of California gnatcatchers. As I walked the trail, I could imagine juvenile gnatcatchers flitting through the park as they looked for a patch of coastal sage scrub of their own.

opment plans for parking lots, day-care centers, a physical plant, and new learning institutes of various disciplines. This is good news, and I will discuss in chapters 6 and 7 why the area along the foothills of the San Gabriel Mountains is so important for California gnatcatcher conservation (box 2.1). But the field station could not escape Mother Nature. In 2013, 6.9 hectares of the field station burned, and in 2017 another 1.6 hectares went up in smoke as I watched from across the street (see box 8.1). Fire is inevitable in coastal sage scrub.[24]

Let me interject a caveat on long-distance gnatcatcher dispersal through landscapes modified by humans. Juveniles find suitable coastal sage scrub patches at random. They are not like bees—they do not receive information from other gnatcatchers as to where these patches are. They are simply setting out and randomly moving through whatever land use they encounter to find a suitable patch before succumbing to predation or exhaustion. Of the one hundred juveniles Bailey and Mock banded in 1998, only twenty-eight were resighted. In Galvin's 1998 study, just twelve of the thirty-eight banded juveniles were ever found again. The low rate of resightings for banded juveniles could be due to too few observers spread across too large of an area to make resightings likely but could also be the result of the low success that juveniles have in finding suitable breeding patches at random.

Diet

If a bird were called a gnatcatcher, would you expect it to eat gnats? This name truly perplexed Robert Woods. Gnatcatchers will in fact eat anything from gnat-sized insects up to decent-sized stick insects and butterflies, beating larger prey items against branches to stun them first before eating. In a 1999 banding study, researchers collected fecal samples dropped by young and adult gnatcatchers in their hands during banding (always a pleasure!) and picked the samples apart under a microscope to identify what they had eaten. The most common food items identified were leafhoppers, planthoppers, and spiders, followed by many beetles, wasps, bees, and ants. Although butterflies and moths were rarely found in the fecal samples—their soft parts are less likely to survive digestion—these insects are also common items on the gnatcatcher menu. I often saw parents head-ing to nests with butterflies hanging out of the sides of their beaks. Parents routinely feed their young the biggest food items they can find, sometimes shoving entire grasshoppers and crickets down their chicks' gullets. Bringing fewer but bigger prey items minimizes the number of trips to the nest, reducing the chance that youngsters will be found by predators.[25]

Birds with insect-based diets are at a disadvantage in urban areas. This is partly due to humans' predilection for killing most insects with chemicals and partly due to the lack of suitable habitats for most kinds of insects in cities. Birds that eat seeds do well in neighborhoods with stocked bird feeders, as do birds that get by on our scraps, such as gulls and crows. Despite this, California gnatcatchers are still able to produce offspring in habitat patches in urbanized areas.[26]

Habitat and Distribution

Although the California gnatcatcher's primary habitat is coastal sage scrub, gnatcatchers are picky when it comes to other habitat characteristics. Gnatcatchers generally avoid steep slopes with greater than a 25 percent rise as well as patches located higher than 500 meters above sea level or too far east where winter temperatures are cold. As a result, there will always be a considerable amount of coastal sage scrub in the region that remains unoccupied by California gnatcatchers because it does not meet their needs—at least until climate change increases temperatures in winter and at high elevation.[27]

Furthermore, California gnatcatchers are choosy about sage scrub even in low-elevation coastal areas. Botanists have identified three or four associations of coastal sage scrub, each characterized by a specialized set of plant species driven by soil type and other site-level environmental factors. Coastal sage scrub in Ventura County is typically dominated by black and purple sage (*Salvia mellifera* and *Salvia*

Box 2.2. Frank G. Bonelli Regional Park, San Dimas,
Los Angeles County, February 12, 2017

My son, the dog, and I searched for California gnatcatchers at this very busy park on a beautiful Sunday morning. We found a female gnatcatcher that we had flushed from a California sagebrush; she called a few times and then disappeared into buckwheat. I was quite surprised to find her there because that patch of coastal sage scrub was dominated by cactus. I had a memory flash of warm spring mornings twenty years before, walking through a field site at Marine Corps Base Camp Pendleton, where a large population of coastal cactus wrens enjoyed the mix of sagebrush and cactus in almost equal proportion. My attention was snapped back to the present by the "churr-churr-churr-churr-churr" of a cactus wren a dozen meters away from where the female gnatcatcher had disappeared. Déjà vu.

leucophylla, respectively). California gnatcatchers tend to avoid these sage scrub associations even though the coastal climate is suitable for them. Black sage produces an especially potent insecticide to protect itself from being eaten, which reduces the abundance of insects that gnatcatchers could eat in that plant community. The California gnatcatcher's avoidance of certain coastal sage scrub associations suggests that it is not a good surrogate species for conserving species that prefer these sage scrub types (box 2.2).[28]

Robert Woods described the California gnatcatcher as "one of the most restricted in distribution of all birds in the United States." The distribution range of the coastal California gnatcatcher subspecies that is listed under the ESA includes six California counties in the United States (Ventura, Los Angeles, Orange, Riverside, San Bernar-

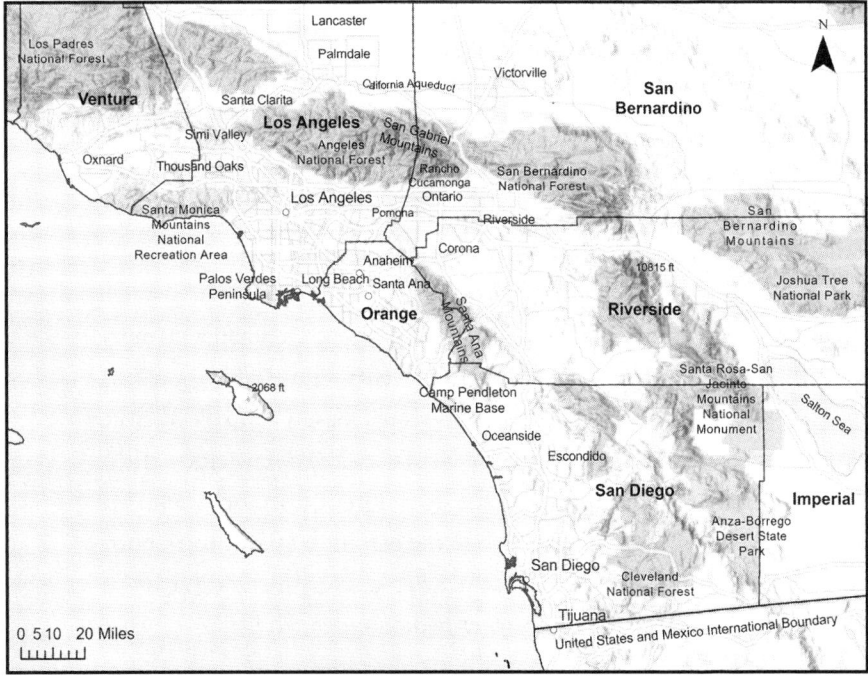

*Figure 2.2. The Southern California region of valleys
circumscribed by mountain ranges.*
(Service layer credits: Esri, HERE, Garmin, Intermap, INCREMENT P, GEBCO, USGS,
FAO, NPS, NRCAN, GeoBase, IGN, Kadaster NL, Ordnance Survey, METI,
© OpenStreetMap contributors, GIS User Community. Map prepared by Jessica Alger.)

dino, and San Diego), plus the northern part of the Baja California
Peninsula in Mexico above 30°N latitude. The species' distribution is
bounded by several mountain ranges to the north and east, too much
black and purple sage to the north, and cold winter temperatures inland (fig. 2.2). Gnatcatchers can't get over the tops of these mountain
ranges, but they can disperse around them through the vegetation at
their base. California gnatcatchers regularly use the foothills along
the San Gabriel Mountains from San Bernardino County to Ventura
County, the foothills of the Santa Ana Mountains running north

to south between Orange and Riverside/San Bernardino Counties, and the base of the transverse Santa Monica Mountains, which run east to west and separate Los Angeles County from Ventura County (fig. 2.2).[29]

The lack of large, stable populations of California gnatcatchers in the eastern inland valleys, despite available coastal sage scrub, is thought to be due to colder winter temperatures in the more desert-like climate. However, with few exceptions, the inland populations of gnatcatchers have not been as systematically surveyed or studied as coastal populations, particularly outside of protected areas, and thus there may be more inland gnatcatchers than currently known. Persistent populations have been observed for decades in Riverside County, such as at the Lake Mathews reservoir where Bill Wirtz and I had a study plot in the early 1990s. Kenneth Weaver found breeding pairs of both California and black-tailed gnatcatchers in the transition zone between coastal sage scrub and desert scrub in Aguanga in Riverside County, and Joseph Grinnell found both species in Palm Springs a bit farther east. With climate change, these inland populations may increase as winter and nighttime temperatures warm (see chapter 3).[30]

Since California gnatcatchers don't occupy every available patch of coastal sage scrub, they exist in many scattered populations. Some of these populations are in core habitat areas that are continuously occupied, and others are in satellite patches that host gnatcatchers sporadically. Habitat patches that are large enough, and of sufficient quality, are source habitats where permanent gnatcatcher populations can produce extra young to bolster or grow the total population of gnatcatchers. Small, isolated habitat patches of poor quality are sink habitats, where gnatcatchers either cannot breed at all or produce few offspring. The metapopulation dynamics of gnatcatchers, whether their populations are stable, increasing, or decreasing, are highly dependent on the ratio of source to sink habitat. If we want California gnatcatchers and other conservation targets to persist in the region, we need to make sure that we have far more source habitat than sinks.[31]

Other factors that drive the metapopulation dynamics of California gnatcatchers include the complex topography of the region, elevation and temperature constraints, winter precipitation, urbanization patterns from neighborhood to landscape scales, and the history of fires and floods that create a diverse and patchy habitat mosaic. Although conservation of the California gnatcatcher is focused on total population numbers north of the U.S.-Mexico border, the trends in source and sink populations around the region serve as one measure of the effectiveness of conservation efforts under the NCCP policy.[32]

3

Population Trends and Current Threats

Despite the innate empathy of its practitioners, conservation biology is a cold, quantitative endeavor. The risk of extinction that a species faces is calculated by numbers of individuals and effective population size—the number of breeding individuals that produce offspring, an amount sometimes much lower than the total population size. Population estimates for the California gnatcatcher are usually reported as breeding pairs. All of those breeding pairs and the new gnatcatchers they produce need to live somewhere; thus, available habitat is also an important quantitative factor. When there are more gnatcatchers than can reasonably live in the available habitat, we know that there will be population declines in the near future if more habitat isn't restored. On the other hand, when populations decline in regions with no change in available habitat, then the cause of the decline may be from other factors, such as weather, disease, or predation. Although multiple factors can drive populations lower, habitat loss is the primary driver of the California gnatcatcher's decline. For that reason, the California gnatcatcher's regional population must be discussed in concert with available habitat.

Population Trends

When California gnatcatchers first gained the attention of naturalists and ornithologists more than a hundred years ago, the species was routinely described as common where found but not found everywhere. For example, in 1898, Joseph Grinnell stated, "Common resident in a few limited localities on brushy mesas and washes, principally along the base of the foot-hills. Numerous in San Fernando Valley and about Pomona and Claremont, but around Pasadena, which is between these two localities and apparently offers similar attractions, I have never seen but one specimen."[1]

About fifty years later, Grinnell and Alden Miller described the species' distribution as "common locally; areas of suitable habitat somewhat reduced within the last twenty years," with individuals observed in six Southern California counties: Los Angeles, Orange, Riverside, San Bernardino, San Diego, and Ventura. There were no efforts at that time to determine or estimate the regional population of California gnatcatchers. Concern over their numbers started in the 1960s, when ornithologists noticed their habitat disappearing more quickly.[2]

In 1980, Jon Atwood estimated between 1,000 and 1,500 breeding pairs, with the majority in Orange, Riverside, and San Diego Counties. In 1992, using more detailed methods, he revised his estimate upward to 1,811 to 2,291 breeding pairs. In the time between these two estimates, Atwood had revisited his fifty-six study sites from 1980 and found that over half had been either destroyed completely (32 percent) or dramatically degraded by construction (27 percent). The broad swath of California gnatcatchers he found across the southern foothills of the San Gabriel Mountains in Los Angeles and San Bernardino Counties was gone. Therefore, his 1980 estimate was likely a substantial underestimate of the actual number of breeding pairs.[3]

In the early 1990s, the Building Industry Association of Southern California hired a contractor to provide an independent estimate of California gnatcatcher populations. The contractor conducted spot surveys in Orange, San Diego, and Riverside Counties to estimate population density per area and then multiplied that number by the amount of available habitat in the region. Ironically, the contractor's total population estimate was lower than Jon Atwood's, around 1,645 to 1,880 breeding pairs. Averaging across all of these studies, we can say with some confidence the total U.S. population was somewhere between 1,600 and 2,300 pairs when the gnatcatcher was listed under the ESA in 1993.[4]

In 1996, the U.S. Fish and Wildlife Service (USFWS) recalculated its estimates of the gnatcatcher population based on the Atwood numbers at the time of listing in 1993, correcting for surveys that had been done afterward. Its estimates increased the gnatcatcher's U.S. population to 3,430 pairs, with the bulk of the population in Orange (908 pairs), San Diego (2,106 pairs), and Riverside (344 pairs) Counties. Small populations were found in Los Angeles County (60 pairs) and San Bernardino and Ventura Counties (12 pairs combined). In 1999, the USFWS revised its estimate upward to almost 5,000 pairs, aggregated from a variety of different surveys and methods across the gnatcatcher's range. Gerald Braden, a biologist at the San Bernardino County Museum, argued that 5,000 pairs was likely a gross overestimate of the actual population. The USFWS never reported another population estimate that high, and 5,000 pairs remains the highest number of California gnatcatchers north of the U.S. border ever reported.[5]

A robust, rangewide population survey remained unavailable when the USFWS released its five-year report for the California gnatcatcher in 2010. Only one large study had been completed: a survey of Orange and San Diego Counties suggested a population of 1,324 breeding pairs in those two counties based on the 44,923 hectares that surveyors were able to access. In 2014, the USFWS estimated the U.S. gnatcatcher population at about 2,500 pairs, and up to 2,900 pairs in 2016. Some of the uncertainty in these pair numbers is due to lack of proper rangewide surveys, but some is due to the gnatcatcher population's response to fluctuating environmental conditions, such as wet winters. Without rangewide surveys, it is impossible to know whether these population fluctuations reflect real changes in the number of gnatcatchers or are caused by survey and estimate error.[6]

Why is it so difficult to survey California gnatcatchers across their U.S. range? These birds are fairly sedentary, sing year-round, and live in a region with bountiful roads and gas stations. Much of the survey problem results from the lack of coordination across studies and inadequate funding for a regional survey effort. In the USFWS's 2016 decision to maintain the gnatcatcher's listing, the service listed the outcomes of available surveys: eleven out of twenty-nine successful nesting attempts in 2014 in the Western Riverside County NCCP planning area; 155 territories occupied by 122 pairs and 33 males in the city of Carlsbad in 2013; habitat occupancy rates in 2012 of 12.7 percent and 34.3 percent in the Central and Coastal reserve areas of the Orange County NCCP plans, respectively; and 436 occupied habitat sites supporting 283 pairs, 122 males, 31 family groups, and 53 transient gnatcatchers on Marine Corps Base Camp Pendleton in 2014. Collating these multitudes of survey measures into a simple estimate of number of pairs is impossible.[7]

Finally, in 2008, a large number of federal and state agencies (headed by the U.S. Geological Survey) joined with local municipalities and nonprofit environmental groups in the San Diego Management and Monitoring Program, supported by considerable funding from SANDAG. As part of the management activities, in 2016 the program initiated a standardized gnatcatcher survey across large areas of public and private property that had been enrolled in conservation programs in Orange and San Diego Counties. The survey included 180 randomly selected points in suitable gnatcatcher habitat across the U.S. distribution range and will be repeated in 2020 and 2024. In the 2016 effort, surveyors observed California gnatcatchers on about a quarter of all sites surveyed.[8]

Although standardized, rangewide gnatcatcher surveys are unavailable, we can carefully use two standardized annual bird surveys for information on rangewide population trends, if not total population size: the North American Breeding Bird Survey and the Christmas Bird Count. These surveys are completed each year by volunteer birders throughout the country following a strict protocol. The Breeding Bird Survey will not detect many gnatcatchers because it is conducted along roads and uses point counts, and both of these approaches are known to miss gnatcatchers. But each survey is conducted along the same route year after year, and the dataset provides a good estimate of species trends across urban landscapes with dense road networks. On the other hand, the Christmas Bird Count calls for birdwatchers to count every bird seen or heard within a 24-kilometer radius, and the survey route often includes stops at natural areas and birding hotspots. Thus, the Christmas Bird Count is far more likely to pick up California gnatcatchers than the Breeding Bird Survey, but it is biased toward protected areas and more natural habitats and comes with many methodological caveats about using it for population trends.[9]

Comparing these two datasets (fig. 3.1), we might surmise that since the 1980s, gnatcatcher populations have increased slightly where there are protected areas or open space with habitat, as observed by the Christmas Bird Count. But across Southern California's urbanizing landscape, gnatcatchers have generally become harder to find, as we might interpret from the increasing number of years with no gnatcatchers observed during the Breeding Bird Survey. The Christmas Bird Count data also illustrate substantial year-to-year fluctuations in gnatcatcher numbers, as expected for a species with a short life-span and high fecundity. In particular, there is a spike in the later

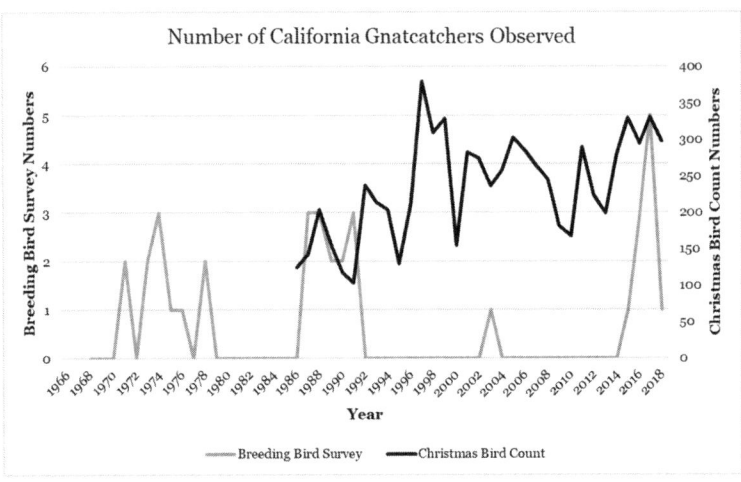

Figure 3.1. The number of California gnatcatchers reported during Breeding Bird Surveys (BBS) and Christmas Bird Counts (CBC) in California, 1966–2018. Due to methodological differences, California gnatcatchers are less likely to be detected in the BBS. The CBC data illustrate how variable gnatcatcher populations tend to be, complicating estimates of their total population size.
(BBS data from the U.S. Geological Survey, Patuxent Wildlife Research Center, and CBC data from the National Audubon Society.)

1990s when almost four hundred gnatcatchers were recorded, reflecting the bumper crop of gnatcatchers produced after the wet winters of the early 1990s as observed in locales such as the Palos Verdes Peninsula. High population variability allows the gnatcatcher population to recover quickly from years that are either too wet or too dry, but high variability also increases the species' extinction risk if consecutive years of bad weather begin to pile up.[10]

Another great source of data is the eBird database, managed by the Cornell Laboratory of Ornithology. This database contains bird observations by tens of thousands of hobbyist and professional birders from around the world, reporting date, time, location, and species. It goes as far back in time as birders are willing to comb through their paper records and enter sightings. (I am still sifting through and entering records from twenty years ago.) The data aren't sufficient to estimate trends over time because observation effort and methodology aren't standardized. Also, the data are skewed to favor those places that birdwatchers like to go, which are usually places that are likely to support many birds and are within driving distance of their home or work. Despite this, we can use these regionwide data to identify where gnatcatchers are consistently seen versus the one-off observations that might be a dispersing juvenile or an overlooked birding hotspot.[11]

The eBird observations from 1900 to the present provide a decent reflection of the large, persistent populations of California gnatcatchers along the coast, as well as smaller populations and sporadic sightings elsewhere. Birdwatchers have clearly identified the large, stable populations of gnatcatchers in San Diego and Orange Counties (fig. 3.2). The large, empty spot in northern San Diego County is Camp Pendleton. Although there is a large and stable population of California gnatcatchers at the base, the area is not open to the public, and therefore the eBird database reflects only those few birders who have access to the base and report their observations. Also evident are the small but stable populations in southwestern Riverside County and in Los Angeles County on the Palos Verdes Peninsula and near Pomona, as well as the smaller but possibly increasing population in Ventura County. The once-common populations of California gnatcatchers in Claremont, as well as elsewhere along the southern foothills of the San Gabriel Mountains, are almost completely absent. On the other hand, Crystal Cove in Orange County, an area that Jon Atwood described as unused in the early 1990s, has now become a reliable place to observe gnatcatchers year-round.[12]

Let's drill down into the areas that play a significant role in the long-term conservation of the California gnatcatcher.

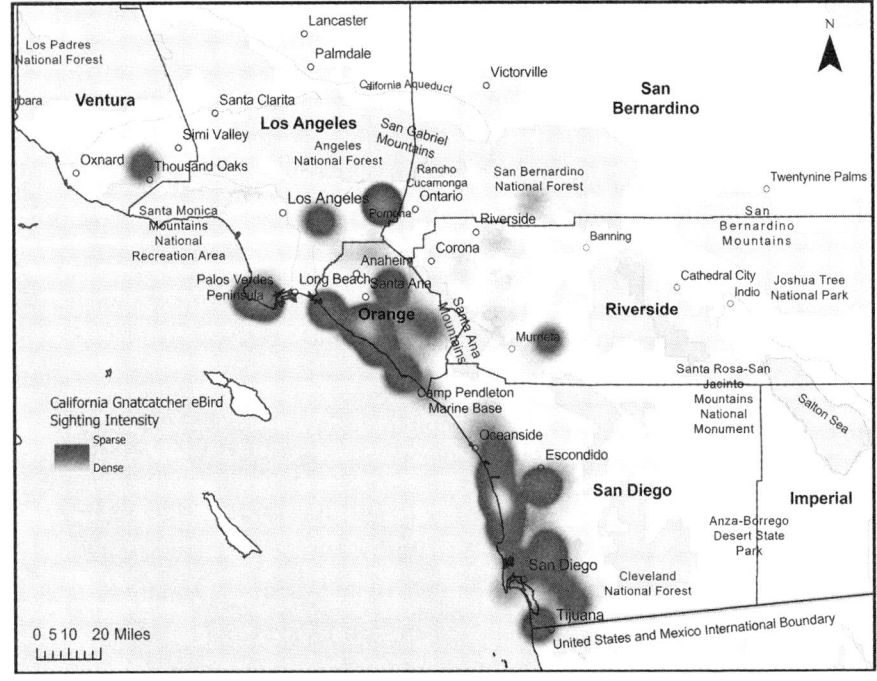

Figure 3.2. A sighting intensity map for California gnatcatchers in the United States using eBird data from 1900 through 2019. Large, stable populations are evident along the coast in San Diego and Orange Counties, as well as the smaller, stable populations on the Palos Verdes Peninsula and at the Frank G. Bonelli Regional Park in San Dimas. There are fewer sightings of gnatcatchers in the counties of Riverside, San Bernardino, and Ventura, due to either fewer birds or fewer birdwatchers there. Marine Corps Base Camp Pendleton appears empty because the base is closed to birdwatchers.

(Data from eBird.com; Basemap source: National Land Cover Database, 2011; Historical Land-Cover Change and Land-Use Conversions Global Dataset [1970, 1990]; UI-UC/ATMO, Department of Atmospheric Sciences, University of Illinois at Urbana-Champaign. Map prepared by Jessica Alger.)

Ventura County: Ventura County is at the extreme north end of the California gnatcatcher's distribution. At the time that the gnatcatcher went through the ESA listing process in the 1990s, ecologists believed that there were no permanent populations of gnatcatchers this far north. In 2001, Atwood and David Bontrager reported four pairs, but no "core" populations of thirty or more pairs. Recently, Daniel Cooper and others reviewed historical and current museum collections, surveys, and eBird records and found that a core population of gnatcatchers has persisted in Ventura County for decades. California gnatcatchers have also been sighted farther north in Santa Paula, as well as in the San Fernando Valley, which connects Los Angeles and Ventura Counties. When I visited one spot in Thousand Oaks in 2017, I found a pair of gnatcatchers and one or more juveniles almost immediately (box 3.1). The passageway through the San Fernando Valley northward into Ventura County will become vitally important if climate change pushes gnatcatchers farther north.[13]

Palos Verdes Peninsula: California gnatcatchers on the Palos Verdes Peninsula (Rancho Palos Verdes) occupy the westernmost edge of the species' distribution. The area's large patches of coastal sage scrub protected in state and county parks are collectively separated from other coastal sage scrub by the vast urban sprawl of Los Angeles — precisely the situation that the NCCP policy was intended to prevent. Although this is considered a stable, core population of at least thirty pairs, the Palos Verdes population demonstrates the species' known vulnerability to bad weather events. For example, the population declined by half after the wet winter of 1993–1994, highlighting the precariousness of its existence on the peninsula. Half of the Palos Verdes gnatcatchers are found in three coastal reserves, which also makes the population vulnerable should a fire wipe out one or two of the reserves at the same time. The remaining gnatcatchers in Los Angeles County are found in scattered small patches to the east.[14]

Box 3.1. California Lutheran University (CLU),
Thousand Oaks, Ventura County, May 21, 2017

From the parking lot, I easily spotted the massive white CLU sign on the hill; California gnatcatchers had been reported there a few weeks before on eBird. As I hiked to within 10 meters of the sign, one male gnatcatcher popped up and called loudly. He then started to dive bomb a pair of cactus wrens in his territory with the ferocity of a chihuahua nipping after a golden retriever. I hiked up around the top of the hill and saw two gnatcatchers chasing off at least two others, moving too quickly for me to get an accurate count. However, from their behavior I assumed that it was an adult pair trying to chase off older fledglings that were stubbornly refusing to leave. As I dodged thick patches of cactus on a narrow path along the ridge, I wondered if I would be able to follow the ridge to where I saw more prime gnatcatcher habitat. No chance. Several new, gated estates blocked my way. Caught behind a gated street, I had to thread my way back along a lengthy fence to return to the CLU sign. That was a lucky break, as I ran into the female gnatcatcher that was probably the mate of the first male I saw. The small open space seemed to support enough habitat for at least two pairs, and it was easily accessible. I wondered if this small population had simply been overlooked by birders until recently or if gnatcatchers had lately moved into the patch. This was one among many sites in Ventura County where eBird records consistently reported California gnatcatchers, and it made me wonder how much our ignorance of this species' northern boundary needlessly complicates its conservation.

Orange and San Diego Counties: The vast majority of California gnatcatchers, as well as the largest and most permanent populations, are in Orange County and western San Diego County. Aiding the persistence of these populations is the massive Marine Corps base at Camp Pendleton in northern San Diego County—506 square kilometers of diverse habitat types, including large stands of coastal sage scrub, supporting a variety of federally listed species. Camp Pendleton was one of the field sites in the early 1990s for our study on the return of California gnatcatchers to an area after fire. The large, stable gnatcatcher population on the base, plus all of the military exercises using flammable objects, made it a fortuitous area to study gnatcatchers and fire.

The persistent core populations in Orange and San Diego Counties factor heavily in all of the conservation plans for the California gnatcatcher. A metapopulation simulation of the Orange County gnatcatcher populations revealed that they are sensitive to variable weather and reproduction rates, with a high risk of extinction over a fifty-year period. Therefore, these core populations along the coast should not be taken for granted.[15]

Santa Ana Mountains/Coyote Hills: Jutting up to the east of Los Angeles, the Santa Ana Mountains and Coyote Hills are vegetated islands in an urban sea, separating the coastal cities in Los Angeles and Orange Counties from the inland valleys of Riverside and San Bernardino Counties. This mountain range creates a rain shadow effect, forcing moist ocean air to rise and cool, dumping its precipitation on the west side of the mountains and creating hotter and drier conditions east of the mountains in the inland valleys. Reports of California gnatcatcher sightings have occurred throughout this chain around its base, from Chino Hills to the north and to Murrieta in the south. The majority of the hills support natural habitat protected by parks and preserves and thus represent a critical habitat area connecting inland and coastal gnatcatcher populations. Researchers and surveyors have consistently overlooked this area.

San Gabriel Foothills (*Azusa, Claremont, Pomona, Rancho Cucamonga*): As described in chapter 2, this area was the epicenter of the species' early life-history research, and even in 1980 it was generally accepted that the area supported at least some gnatcatchers. But since then, ecologists have basically written off this eastern corner of Los Angeles County and western San Bernardino County, probably because of the explosive urban development in the area in the past thirty years (box 3.2). Nevertheless, I haven't given up hope that this area could be restored and rediscovered by gnatcatchers. The eBird database includes observations of gnatcatchers throughout the area, including small, but stable, populations in Rancho Cucamonga and Redlands (fig. 3.2). A recovered population here would connect inland populations in San Bernardino and Riverside Counties with populations in the San Fernando Valley and Ventura County. The area may also be a critical evolutionary hotspot in the region, as evidenced by high genetic variability in a variety of animals, and thus worthwhile to protect for reasons beyond the conservation of current species.[16]

San Bernardino/Redlands: This area is due east of Claremont and Pomona along the foothills. In 1980, Atwood could find no reports of California gnatcatchers observed there after 1960, and in a 1990 survey he found just one bird, around Cajon and Lytle Creek Wash. Atwood still assumed that about fifty pairs were breeding in the county, and there were occasional sightings in the 1990s around the Etiwanda Fan and north Rialto area. As of 1999, Atwood and Bontrager tallied reports of twenty-seven pairs in San Bernardino County. The eBird database suggests there is a considerable population in the Jurupa Hills that warrants protection as a stepping stone, connecting gnatcatcher populations along the San Gabriel Mountains via the

---

Box 3.2. Heritage Park, Rancho Cucamonga,
San Bernardino County, April 30, 2017

Tucked away at the base of the San Gabriel foothills, Heritage Park is one-half baseball diamonds and playground, one-quarter equestrian center, and one-quarter natural area. A pair of birdwatchers reported seeing one California gnatcatcher here in 2013 and one "at Susie and Rob's house" across the street from the park later the same day. Other birders have reported an impressive variety of birds at the park, but none has reported another gnatcatcher. The park looked small on the map, and it seemed even smaller in person. My son and I walked around and through the majority of the 4-hectare natural area in about twenty minutes. The available coastal sage scrub was sandwiched between a residential area and a riparian wash. It was a lovely little patch of sage scrub that a gnatcatcher could call home, but that's it—just one, or maybe one very efficient and experienced pair. However, I wasn't surprised that a California gnatcatcher had passed through that patch, perhaps hanging out for a few weeks to fuel up and look around before setting out for larger patches. The park was a great oasis for California gnatcatchers and many other birds moving through the creeping suburban sprawl along the San Gabriel foothills.

Santa Ana River valley to the populations at Lake Mathews in Riverside County and to populations in the Santa Ana foothills.[17]

Baja California Peninsula, Mexico: No recent, systematic surveys have been conducted anywhere south of the U.S.-Mexico border, but that hasn't impeded wild guesses about population numbers there. When the subspecies was listed as threatened under the ESA in 1993, the USFWS estimated that there were perhaps 2,800 pairs of California gnatcatchers in Mexico. At that time, the building industry of Southern California was arguing that listing the species was unnecessary, claiming there were 1.5 million pairs (or 2.5 million birds) in Mexico. The estimate was based on an unpublished consultant's report written for the building industry, extrapolated from sightings of 396 gnatcatchers. In 2016, the Pacific Legal Foundation, a public interest law firm representing the building industry, sued the USFWS to delist the gnatcatcher based on the estimated millions of California gnatcatchers south of the U.S. border. Some simple calculations can illustrate the impossibility of this number.

In 2016 the USFWS estimated that there were about 689,749 hectares of coastal sage scrub south of the U.S.-Mexico border. If 1.5 million pairs occupied every last bit of 690,000 hectares of available habitat, that would leave less than 0.5 hectare of habitat for each gnatcatcher pair territory. As detailed in chapter 2, a pair of gnatcatchers in the United States needs at least 1 hectare of the most productive coastal habitat for a breeding territory and upward of 18 hectares of poor inland habitat—and not every hectare of coastal sage scrub is going to have a gnatcatcher pair on it. If we use the habitat occupancy rate of 25 percent found during the 2016 survey of the San Diego Management and Monitoring Program and optimistically assume that all of the 690,000 hectares of coastal sage scrub in Mexico is of the highest quality, then the maximum number of California gnatcatchers that there could be in Mexico is around 173,000 pairs. The building industry's million-pair estimates are hard to believe.[18]

Despite the lack of exact numbers for the world's population of California gnatcatchers, conservation efforts in the United States need to move forward with the knowledge on hand. We know that: California gnatcatchers need coastal sage scrub; 60–90 percent of coastal sage scrub has been lost to development; the majority of coastal sage scrub suspected of supporting California gnatcatchers is privately owned; and fragmentation of the remaining habitat will make it difficult for gnatcatchers to find mates and resources. As Atwood warned, "The long-term survival of *P. c. californica* [coastal California gnatcatcher] in the United States would be questionable even if there were currently many thousands of breeding pairs remaining north of Mexico. Without reversal of present land use trends, California gnatcatchers will be extirpated from most or all of their range in the United States in the near future."[19]

Current Threats

If the California gnatcatcher has shared Southern California with humans for thousands of years, why is it faring so badly now?

In its 2016 decision to maintain the listing status of the coastal California gnatcatcher, the USFWS identified habitat loss and fragmentation, grazing in Mexico, and changes in temperature and precipitation driven by climate change as primary causes of the gnatcatcher's endangerment. Other factors made worse by climate change include increased frequency and intensity of wildfires and the conversion of coastal sage scrub to other vegetation types that are better adapted to more frequent fires and drier conditions. These threats work synergistically to decrease numbers of California gnatcatchers and make conservation efforts more challenging. Although brood parasitism and predation are risks to individual gnatcatchers, the USFWS determined that these factors do not imperil the entire species. Each of these causes calls for a different policy prescription for conservation.[20]

According to the USFWS, habitat loss is the most pressing threat against the long-term existence of the California gnatcatcher. Broadly speaking, habitat loss is measured by how much habitat remains, while habitat fragmentation is measured by where the remaining habitat is and the size of the fragments, the kind of land use that isolates those fragments, and how long they've been isolated. Although ecologists treat habitat loss and fragmentation as different phenomena, they are usually concurrent. Habitat loss results when one kind of habitat is converted to another kind of habitat or to a human land use. In the process of habitat loss, small habitat fragments may remain scattered across a landscape. It is possible to have habitat loss without fragmentation if one large patch of habitat remains and also to have habitat fragmentation without habitat loss if a linear barrier is erected through a habitat patch—say, for example, a border wall or a highway. For the California gnatcatcher's conservation, we are concerned about both habitat loss and fragmentation, because some remaining habitat fragments may be too isolated and small for gnatcatchers to find and use.[21]

The largest threat to the California gnatcatcher is the loss of its habitat to human land uses. A century ago, coastal sage scrub in Southern California was lost mainly through conversion to small farms and citrus orchards or to cattle grazing. Since then, habitat loss has been driven by urbanization. When the USFWS listed the coastal California gnatcatcher as threatened under the ESA in 1993, the agency noted that over half of the gnatcatcher's coastal sage scrub habitat had been eliminated in Los Angeles, Orange, and San Diego Counties, within the core distribution of the subspecies. In the service's 2016 review, urbanization remained a top threat in both the United States and Mexico: "Urban development continues to result in the destruction, modification, or curtailment of the coastal California gnatcatcher's habitat, and represents a current, medium-level stressor to the coastal California gnatcatcher across its range in the United States and Mexico that has the potential to result in the loss of gnatcatchers at the population level and the loss of large but isolated patches of habitat. This stressor will continue to impact the subspecies and its habitat into the future."[22]

Between 1945 and 1990, the rate of loss and fragmentation of the California gnatcatcher's habitat varied by county. Habitat loss reached 50 percent in Orange County, 59 percent in Riverside County, and 60 to 65 percent in San Diego County. Fragmentation of coastal sage scrub regionwide increased fourfold from 1931 to 1990. In western Riverside County, suitable habitat losses ranged from 12 percent (pre-1997 to 2003) to 41 percent (2003 to 2012). The conversion of coastal sage scrub to agriculture, cattle grazing, and urbanization at the north end of the Baja California Peninsula reduced the connectivity between the gnatcatcher populations in Southern California and those on the southern half of the Baja California Peninsula. This loss of connectivity reduces gene flow and the ability of California gnatcatchers on the peninsula to shift their ranges north to adapt to climate change.[23]

Luckily for the California gnatcatcher and its conservation, gnatcatchers don't seem to shy away from the edges of their patches. Some species are interior species; they tend to stay toward the center of the patch, particularly if the area surrounding the patch is radically different from the preferred habitat. For example, the federally threatened northern spotted owl (*Strix occidentalis caurina*) is an interior species. These owls need very large stands of old growth forest to breed, making conservation challenging. Other species are edge species, and they actively seek out the edges between two habitats. White-tailed deer (*Odocoileus virginianus*), for example, are most commonly found in the boundary between woodlands and pastures or agricultural fields. But some species, such as the California gnatcatcher, seem to not care either way. They will establish territories anywhere from the middle to the edge of a coastal sage scrub patch and will often use neighboring vegetation to feed, especially riparian areas. This is not surprising, since gnatcatchers evolved in a diverse region frequently affected by fires, where chaparral, coastal sage scrub, riparian areas, and grasslands have always been comingled across the landscape. This also means that conservation plans can include coastal sage scrub patches and corridors through urban neighborhoods, because gnatcatchers could use them to disperse to other patches and find food, even if those patches are too small for breeding.[24]

To reconnect the isolated gnatcatcher populations on the Palos Verdes Peninsula, in Ventura County, and in the Coyote Hills—the spine of foothills running between Orange and Riverside Counties— we need to preserve and restore gnatcatcher habitat across the region. The small inland populations in eastern Los Angeles County, Riverside County, and San Bernardino County are similarly isolated from coastal populations by an urban sea. In these areas, coastal sage scrub has dropped below 10 percent of the landscape and individual gnatcatchers show a loss of genetic diversity, resulting from insufficient connectivity with other populations. California gnatcatchers have successfully reestablished themselves in restored coastal sage scrub on the Palos Verdes Peninsula, suggesting cautious optimism for future habitat restoration efforts. As will be explained in chapter 6, the NCCP policy can link up habitat networks within their planning area, but the program's protected areas need to be coordinated across the region to assure that the habitat patches are connected at the regional scale.[25]

The interior populations in eastern Los Angeles County, western Riverside County, western San Bernardino County, and Ventura County are often overlooked by researchers and conservation agencies, under the assumption that there are few stable populations in these areas—a risky supposition. Until rangewide, systematic breeding surveys are conducted across the California gnatcatcher's range, none of these areas should be written off as unimportant to the gnatcatcher's conservation. With the expected near-future climate changes, these areas might become critically important to the long-term survival of the species.[26]

The conversion of coastal sage scrub to grasslands through cattle grazing is no longer a significant threat to the California gnatcatcher in Southern California, but it remains a significant threat in Mexico. The negative effects of grazing on coastal sage scrub were far more pervasive in the late nineteenth century, when cattle ranching was still common across Southern California. By the early 1900s most ranches had been converted to citrus groves and suburbs, preserved as open space, or otherwise taken out of grazing use by city ordinances. The few ranches that remain today are under intense development pressure. For example, in January 2017, the City of Industry placed a $100 million bid for the 1,010-hectare Tres Hermanos Ranch, just south of Pomona in central Los Angeles County. Perched at the northern end of the Santa Ana Mountains between Orange and Riverside Counties, for years the ranch has been targeted for development. Birdwatchers have reported observations of California gnatcatchers to eBird throughout that chain of foothills, with several observations made just adjacent to the Tres Hermanos Ranch in Powder Canyon and the Firestone Scout reservation. Advocates for open space hoped that the City of Industry's plan for the ranch was to preserve it as open space. But several months after acquiring the ranch, reporters revealed that the city was planning to use the land for a 440-megawatt solar energy facility, large enough to meet 10 percent of the electricity demand of Los Angeles at a time of an overabundance of cheap electricity in the state. As I write this, the plans for the solar facility have stalled while the City of Industry battles two neighboring cities (Chino Hills and Diamond Bar) in six lawsuits over control of the property—all three cities were eyeing the property for housing. According to one member of the environmental group Save Tres Hermanos Ranch, none of the three cities has an appropriate plan for the ranch. He said, "We can't just fill up everything with development. We're going to have to come up with smarter ways to meet our housing needs and preserve our quality of life."[27]

Grazing in Mexico is a different story. Cattle still graze in coastal sage scrub, and coastal sage scrub is cleared and replanted with pasture grass. Thus, according to the USFWS, grazing still remains a high risk to coastal sage scrub and the California gnatcatcher in Mexico.[28]

Direct and indirect climate-change impacts generally increase the risk of extinction for the California gnatcatcher in the long term. Climate change directly affects temperature and precipitation and indirectly affects the probabilities of fire, flooding, or storms that are influenced by changes in temperature and precipitation. Let's take direct and indirect impacts in turn.

Temperature and rainfall directly affect the survival, breeding behavior, and fecundity of California gnatcatchers, and thus climate determines the limits of the species' distribution. Generally, cold overnight temperatures in the winter determine the California gnatcatcher's eastern distribution limits, as well as its elevational limits in the foothills—winter nights are colder in deserts and at higher elevations. Proximity to the Pacific Ocean moderates the climate of coastal habitat: these areas are wetter year-round than inland areas and have cooler summer temperatures and warmer winter temperatures. Precipitation governs the timing of a gnatcatcher's spring plumage molt and the number of eggs laid per clutch, with smaller clutches laid in dry years. Gnatcatcher survival declines during cold, wet winters— "cold" means average January temperatures falling below 2.5°C. Fewer insects are available during cold, wet winters, leading to food shortages when gnatcatchers need to burn more energy to stay warm. However, a string of cold, wet winters often produces a bloom of new vegetation, which provides resources for a bloom of insects once temperatures recover. For example, after several cold, wet winters in the early 1990s, the 1993 winter was warm and dry, creating a bumper crop of gnatcatchers in 1994; multiple studies in Orange County and the Palos Verdes Peninsula observed gnatcatcher population declines before 1993 and increases after 1994. Regional climate patterns drove this synchrony in gnatcatcher population dynamics.[29]

Since the 1950s, the minimum, maximum, and average annual temperatures in Southern California have all increased by 0.5°–1.6°C. Over the next thirty years, temperatures throughout the California gnatcatcher's range are expected to increase by 1.6°–2.5°C, with longer periods of drought punctuated by more intense wet periods. California experienced something similar in the drought of 2011–2016, ended by the record-breaking 2016–2017 winter storms. A bump in gnatcatcher numbers recorded in the Breeding Bird Survey and Christmas Bird Count might reflect a more abundant insect resource after that record winter precipitation (see fig. 3.1).[30]

Long-term changes in temperature and precipitation may benefit some populations of California gnatcatchers and cause others to decline, but without more locally specific models it is difficult to say exactly where these impacts will occur and when. Higher-elevation and inland areas should become more welcoming to gnatcatchers as regional winter and overnight temperatures increase, reducing the energetic strain on gnatcatchers to find food and keep warm. Some plants and animals, such as the endangered Quino checkerspot butterfly (*Euphydryas editha quino*) in Riverside County, are already moving to higher elevations in response to climate change.[31]

Climate change will also have indirect effects on gnatcatchers via changes to their habitat. For example, if coastal sage scrub appears at higher elevations, more California gnatcatchers should be found there if the temperatures are suitable. But climate change may have negative effects on coastal sage scrub, too, hastening its transition to grassland or nonnative plant communities due to altered nutrient cycling or increased fires driven by reduced precipitation. The three factors that are expected to affect coastal sage scrub most directly are a change to wildfire patterns, a transition to grassland due to increased fire, and increased competition from nonnative, invasive plant species. In 2016, the USFWS determined that indirect effects of climate change are medium-level stressors on the long-term persistence of the California gnatcatcher.[32]

As with many Mediterranean-type ecosystems, fire is a natural part of the ecology of coastal sage scrub and, therefore, for the animals associated with it. Wildfires are more common near human activity, particularly in the urban-wildland interface where residential neighborhoods abut natural areas along the foothills. In the United States, fires at the urban-wildland interface can be severe in areas where prescribed burning to manage fuel loads has not been practiced, and during times of hotter, drier conditions.[33]

Fire affects California gnatcatchers directly through increased mortality and habitat loss. Gnatcatchers are not particularly strong fliers, and a hot, rapidly moving fire can easily overtake and kill gnatcatchers trying to flee. Fires that are naturally ignited by lightning tend to burn cooler and slower and present less of a mortality risk. These cooler fires occur mostly in the winter and early spring, before the breeding season, and adults can readily escape these fires. After a fire, gnatcatchers will hang out in the surrounding unburned habitat, and over time they will recolonize the burned patch as the vegetation recovers. The higher the habitat quality, the faster this recolonization takes place. But the unburned refugia patches surrounding a burned area have a temporary overabundance of adults, which may reduce the reproductive capacity of all the pairs in the unburned patch until the excess adults can move back out. Gnatcatchers can recolonize recovering coastal sage scrub three to five years after a fire in areas where the vegetation has sufficiently recovered; although extremely hot, severe fires can delay recolonization for ten years or more when the coastal sage scrub recovers more slowly. Depending on the conditions at the time of the fire and the site's ecological history, some coastal sage scrub stands may never recover, instead transitioning to grassland or becoming dominated by nonnative plant species, precluding gnatcatcher recolonization. Areas that are repeatedly burned are also more likely to convert from coastal sage scrub to grassland and become unusable to gnatcatchers.[34]

In 2016, the USFWS estimated that 45 percent of potential California gnatcatcher habitat in Southern California had burned between 2003 and 2015. Over 20,000 hectares of this percentage had burned more than once, and some patches had burned up to four times in that period. The amount of habitat lost to fires in a year is difficult to overstate—in one of Orange County's conservation planning areas, 20 percent burned in 2006, and another 40 percent burned in 2007. Therefore, fire remains an ever-present challenge to biodiversity conservation in the region. Habitat preserves must include prescribed burning in their management plans, first to control fire risk to the preserves and the areas around it, and second to maintain a diversity of years-postfire patches for California gnatcatchers and other coastal sage scrub species. A string of very active fire seasons over several consecutive years could temporarily eliminate a significant percentage of gnatcatcher habitat, placing the entire regional population at risk. Thus, fire needs to be managed at both the site and regional scale.[35]

The conversion of coastal sage scrub to another habitat type is more likely to occur where invasive species are present, especially in areas with more frequent fires. Patches of gnatcatcher habitat that remain after an area is urbanized, such as in small parks and open space left as a development offset, are burdened by increased fires and competition from invasive plants that escape from gardens and landscaping. Climate change can also affect an area's hydrology and nutrient cycling, favoring invasive plant species over native species. California gnatcatchers will abandon areas that have become dominated by invasive plants, even if some sage scrub species are still growing there. Back in the 1990s, one of our gnatcatcher survey sites at Camp Pendleton was overgrown with 2-meter-high black mustard (*Brassica nigra*). Spanish colonists brought black mustard with them from Europe and planted it to use as a spice. That site was my least favorite to survey, because I was guaranteed to get a head full of spiders and not see a single gnatcatcher for my trouble.[36]

The last threats, and those of least concern according to the USFWS, are the risks posed to juvenile gnatcatchers by predation and brood parasitism. Throughout their range, California gnatcatchers are confronted with both predation by snakes, mammals, and birds and brood parasitism—laying eggs in another's nest—by the brown-headed cowbird (*Molothrus ater*). Brood parasites, predators, and nonnative competitors often thrive in the edges between natural habitat and human land uses, particularly in agricultural or urban areas. When combined, these factors can lead to half of all gnatcatcher nests failing in a year. Despite this, the USFWS determined that neither predation nor brood parasitism is a significant threat to the long-term persistence of the California gnatcatcher, because adults renest so readily after a nest failure.[37]

Globally, there are a few species of birds that have evolved the lifestyle of a brood parasite; the cuckoo family is the classic archetype. Everyone is familiar with the common cuckoo (*Cuculus canorus*), which says "koo-koo!" like the clock and is a brood parasite not found in North America. Interestingly, none of the three species of cuckoos that we do have in North America are brood parasites. One of them is the federally threatened yellow-billed cuckoo (*Coccyzus americanus*) and is the subject of an ESA drama that is similar to the California gnatcatcher's story in this book. Brown-headed cowbirds, however, are brood parasites and native to the continent. Like other brood parasites, the cowbird lays one of its own eggs in the nests of host parents, often smaller songbirds. Cowbird eggs are larger than the hosts' eggs, as are cowbird hatchlings, because cowbird eggs incubate for a shorter time and hatch before the hosts' eggs do. The larger cowbird egg and hatchlings can monopolize resources, such as heat during incubation and food during feedings. Host defenses against nest parasitism include abandoning a nest if a cowbird egg is laid in it, puncturing the cowbird egg to prevent development, or pushing it out of the nest. California gnatcatchers usually opt to abandon the nest. Even with these host adaptations, cowbird brood parasitism has contributed to the decline of several imperiled bird species in North America. In the case of the federally endangered Kirtland's warbler (*Setophaga kirtlandii*), brown-headed cowbird eradication programs in its breeding range allowed the species to make a full recovery—the species was officially delisted on October 9, 2019.[38]

The earliest evidence of brown-headed cowbird parasitism of a California gnatcatcher nest is from an egg set collected in 1933, preserved at the San Bernardino County Museum in the city of San Bernardino. Parasitized gnatcatcher nests that aren't abandoned always fail. Cowbird eggs are roughly twice the size of gnatcatcher eggs and cowbird chicks are even larger; the gnatcatcher's young don't stand a chance. Despite this, nest parasitism by brown-headed cowbirds is not a significant cause of nest failure for California gnatcatchers overall, since the breeding season for gnatcatchers is longer than that of cowbirds. That said, many conservation plans for the California gnatcatcher list cowbird mitigation programs among the possible conservation activities, which may help small populations that are struggling under such other stresses as increased fire or vegetation type conversion.[39]

Although nest abandonment due to parasitism is a cause of nest failure, predation of the eggs and young poses a greater risk. The relief from brood parasitism that gnatcatchers get during the cowbirds' off-season is negated by constant and intense predation pressure throughout the gnatcatcher's breeding season. The California gnatcatcher happens to live in a particularly predator-rich environment, and the rates of nest predation that it suffers are considerably higher than the average for birds laying eggs in open-cup nests. The fragmentation of coastal sage scrub into patches surrounded by human-dominated land uses increases the abundance and kinds of predators in the landscape. There are many predators of gnatcatcher eggs and young, including lizards, snakes, ground squirrels, mice, coyotes, weasels, other birds such as shrikes, roadrunners, and wrens—and, more recently, invasive Argentine ants. But California gnatcatchers evolved in this environment; they have been dealing with most of these nest predators for over ten thousand years. Their ability to build and fill new nests quickly allows them to persist in the region.[40]

In sum, although the most daunting threat to California gnatcatchers in Southern California is the loss of their habitat to urban development, other factors would also be issues even if all development stopped today. The remaining coastal sage scrub, protected or not, must be sufficient to help California gnatcatchers adapt to increased temperatures, fires, invasive species, and urban-loving predators. California gnatcatchers face similar threats in Mexico as in the United States, and therefore we cannot assume that faltering gnatcatcher populations in Southern California can be rescued by a wave of dispersing gnatcatchers from Mexico.[41]

4

California Gnatcatcher Taxonomy

The decades of controversy over whether California gnatcatchers in Southern California are suitably distinct from California gnatcatchers elsewhere, and thus merit protection under the Endangered Species Act, are not unique to the California gnatcatcher: challenges to taxonomic distinctiveness is a common approach to elicit a delisting decision. Some of this controversy reflects the complexity and arbitrary nature of taxonomic science and its protocols for what we consider definable units such as *species* and *subspecies*. Uncertainty in species boundaries and evolutionary uniqueness provides fuel for countless lawsuits over ESA listings, including lawsuits over the listing of the coastal California gnatcatcher.

The California gnatcatcher's taxonomic status has an oversized influence on conservation policy in Southern California because species classification is complicated. Species classification is complicated because evolution isn't orderly or linear. The evolutionary process doesn't start with a figurative tree trunk, then neatly expand into clearly separate branches and twigs. Instead, it is a gnarly bush with the capacity to graft nearby branches and twigs back together, often repeatedly. The threshold where two species are so different that they can no longer breed and produce offspring can be porous and context-dependent. As the cost of sequencing entire genomes has declined, we have gained more information on species' genetics and evolution, particularly on the many mechanisms of gene flow between species. The way we classify species doesn't necessarily reflect the amount and direction of gene flow among species. This is especially true for birds. Since the lines of genetic differentiation between species are not always clear, creating family trees for birds as they evolved from their dinosaur ancestors has been an ongoing headache. The maturing field of phylogenomics has revealed surprising insights regarding the evolutionary relationships among different groups of birds and the order in which certain adaptations arose in bird lineages, such as nocturnal behavior and the loss of flight.[1]

The evolutionary history of birds contains a great deal of hybridization and back-crossing across what humans would consider species boundaries. For example, the familiar mallard (*Anas platyrhynchos*) with the distinctive green head hybridizes with more than fifty other species of ducks. Evolution may even favor birds with a high tolerance for breeding across these boundaries. But the cosmopolitan nature of avian breeding and evolution creates enforcement problems for species-focused laws such as the ESA. For example, when species listed under the ESA hybridize with common, unprotected species, the offspring aren't automatically listed, too; protection for hybrids is decided by the U.S. Fish and Wildlife Service or the National Marine Fisheries Service on a case-by-case basis. If hybridization helps prevent the loss of an endangered species' genetic uniqueness, hybrids are protected. But if hybridization threatens to genetically overwhelm a listed species, hybrids are treated as a threat and eliminated as part of the listed species' recovery plan.[2]

While uncertainty over species boundaries is an interesting debate among scientists, uncertainty directly influences policy decisions for biodiversity protection. The ESA was originally written and passed before genetic techniques became widely used for taxonomy and conservation, and the act's amendments have not kept pace with the science. As with many environmental laws, executive branch agencies, such as the USFWS, have considerable discretion over how new data will be incorporated into regulatory enforcement. Species can lose their legal protection if new science demonstrates that the listed unit is not suitably different from other species or subspecies. In these cases, either the listing agency decides to delist the species based on new taxonomic data or the agency is forced to delist the species by a successful lawsuit. The loss of legal protection due to new research is not unique to species protection laws in the United States. Recently, the Western Australian government sought to declare the dingo a non-native species and remove its legal protection, based on a 2017 scientific paper arguing that dingoes (*Canis dingo*) were not genetically or morphologically distinct from domestic dogs (*Canis familiaris*). If genetics can downgrade dingoes to dogs, then many other species can lose their claim to uniqueness, too.[3]

What Is a Species?

Humans prefer to place things into neat categories with clear boundaries: masculine or feminine, circle or sphere, Chicago Cubs or White Sox fans. But as nature constantly reminds us, most things cannot be neatly categorized. Consider these three examples: gender, dimensions, and baseball. In humans, the construct of gender categories is highly variable within and across cultures; many cultures past and present have recognized more than two gender categories and the considerable overlap among them. In geometry, some objects and patterns cannot be categorized into one dimension (a line) or two (a circle) or three (a sphere). For these objects, fractal mathematics comes in handy, revealing a world of "between" dimensions with shapes that are neither lines nor circles nor spheres. When the Chicago Cubs won the World Series in 2016 after a 108-year championship drought, Cubbies fans found out that many White Sox fans had secret blue shirts tucked away in their closets.[4]

Efforts to categorize organisms into species encounter the same boundary problem. The system for sorting organisms into categories, called taxonomy, dates back to Carl Linnaeus in 1735, one hundred years before Charles Darwin's *On the Origin of Species* and Gregor Mendel's insights into the existence of genetic material. This classification system depends on organisms' similarity to one another. Organisms are categorized into increasingly specific groups, and each organism can only belong to one group in each layer: Kingdom, Phylum (or Division for plants), Class, Order, Family, Genus, Species. In this book, where I first discuss a species or subspecies in the context of the ESA or taxonomy, I usually provide the scientific name next to the common name—the two- or three-word Latinized name that scientists use to reduce confusion across languages or regions. The first two words are the genus and species, respectively; for example, *Polioptila californica* is the California gnatcatcher. A third term for animals—*Polioptila californica californica*—identifies subspecies, in this case the coastal California gnatcatcher, a subspecies of California gnatcatcher that lives in Southern California and northern Mexico. Subspecies is the rank below species and indicates two or more populations of a species that are sufficiently different to merit distinction; there can never be just one subspecies in a species. Over time, subspecies may differentiate from other subspecies and become full species. Not all scientists subscribe to the concept of subspecies.[5]

There are many different ways to define species—these are called species concepts. Perhaps the best known is the *biological species concept,* which places organisms in two different species if no fertile offspring can be produced. A well-known example is the mating of a domestic horse (*Equus ferus caballus*) to a domestic donkey (*Equus africanus asinus*), producing a mule, which is always sterile. Since mules do not have the ability to self-propagate, they have not been assigned a unique scientific name. As you can see from their scientific names, horses and donkeys are placed in the same genus but different species. This species concept is fairly straightforward to use if you can try to mate two organisms of the opposite sex.[6]

But the biological species concept is not straightforward to apply in the wild. Two species may not produce offspring because they never encounter each other in nature—and if they did, their hybrid offspring might be fertile. Species can be physically isolated by the time of day that they are active or by a mountain range or an ocean. In other cases, their songs or behavior are so different that they don't recognize each other as potential mates—a common driver of speciation in birds. After being separated by the Great Plains of North America for millennia, the barred owl (*Strix varia*) has expanded its range westward into the range of endangered spotted owls (*Strix occidentalis*). Barred and spotted owls now interbreed and produce fertile hybrid offspring, greatly complicating spotted owl conservation. In this case, the USFWS has determined that the hybrid offspring of barred and spotted owls should not be protected under the ESA, over fears that the barred owl hybrid offspring will eventually eliminate spotted owls through interbreeding.[7]

Due to the shortcomings of the biological species concept, taxonomists have developed a suite of other concepts, such as the *ecological species concept, evolutionary species concept,* and *phylogenetic species concept.* Everyone has their favorite, usually the one that best fits the organisms with which they work. Because of the use of many different species concepts, scientists can disagree on species' boundaries even when looking at the same data. As you can imagine, this diversity of perspectives doesn't make it easy to enforce the ESA and other laws that depend on taxonomic clarity.[8]

The development of new species can occur through adaptation

to the environment or through genetic mutations, but usually both mechanisms are involved. Speciation driven solely by environmental factors tends to be slower than when driven by genetics, and genetic differences between species can be incredibly small. For example, just one gene determines whether a chinook salmon (*Oncorhynchus tshawytscha*) conducts its breeding run in the spring or the fall, effectively separating the species into two distinct populations—this separation forms the basis of a petition to list the quickly declining spring run in the Klamath River under the ESA. If this separation persists, these two populations may eventually become two species with DNA too different to combine.[9]

Thus, identifying species and subspecies, and determining which organisms belong in which categories, is a far more complex process than most people realize, including policy makers. Scientists recognize the fuzziness and somewhat arbitrary nature of the boundaries around what we call a species or a subspecies. Ornithologists, in particular, revel in decades-long taxonomic debates over bird classifications and species relationships. But taxonomic decisions made by scientists have legal implications that directly affect an individual organism's protected status. The debate over the taxonomic classification of California gnatcatchers has lasted for over a century, complicating the legal process for its conservation.[10]

1881–1926: *Polioptila californica* Is Declared a New Species

While genetic analyses have become a more common tool used to differentiate species, we still use morphology (what an organism looks like on the outside), physiology (what it looks like on the inside), behavior, and geographic range to identify species because this information is less costly and more readily available. In 1881, the California gnatcatcher's first recognition as a full species was based on the information available at the time: morphological and behavioral characteristics, and geographic range.

William Brewster, a naturalist from the Northeast, was an expert at the use of morphological differences to define new species. He was the first to elevate the California gnatcatcher to species level by differentiating it from other southwestern gnatcatcher species known at the time: *P. caerulea,* the blue-gray gnatcatcher found throughout Canada, the United States, Mexico, and the Caribbean; and *P. melanura* (then called *P. plumbea*), the black-tailed gnatcatcher in the U.S. Southwest and Mexico. The "*P.*" is shorthand for *Polioptila,* the name of the gnatcatcher genus. Brewster differentiated these gnatcatcher species using differences in plumage color. This was a very common technique at the time, given that most naturalists shot first and examined the dead specimens later in a museum. These plumage descriptions were incredibly detailed, almost feather by feather, with color distinctions that would put a paint store to shame. In 1881, Brewster wrote:

> Upon comparing the California bird with [*P. melanura*] as represented by my Arizona specimens, the following differences appear. The ash of the upper parts is decidedly plumbeous instead of bluish; the throat, breast and sides dull ashy instead of ashy-white; the abdomen, crissum and under tail-coverts fulvous, in some specimens pale chestnut; the light edging of the tail feathers confined to the outer pair of rectrices (with sometimes a slight tipping on the second pair) and on these restricted to the extreme tips and a narrow margin along the outer web; the lining of the wings pearl-ash instead of white and the secondaries and tertials edged with light brown. There is no pure white anywhere on the bird, and the general aspect beneath is nearly as dark as in the Catbird.[11]

Naturalists at the time also took pains to measure every part of their specimens, using body measurements, such as leg length and bill width, as additional traits to differentiate species. Brewster noted longer, thinner bills and legs in California gnatcatchers than in black-tailed gnatcatchers, but shorter wings and tails. Neither whiteness of feather edges nor morphological features are evident through binoculars on a 12-centimeter bird in the field. Brewster named this species of gnatcatcher *Polioptila californica,* described as the "California black-capped gnatcatcher," which was later shortened to California gnatcatcher. Brewster's taxonomic work also highlighted that California gnatcatchers can co-occur with black-tailed gnatcatchers at the eastern edge of the California gnatcatcher's range, and the plumage differences between them in that zone are quite obvious.

Thus, as the nineteenth century gave way to the twentieth, the California gnatcatcher was considered a full species called *Polioptila californica,* inhabiting a thin range along the coast of Southern California and southward into the Baja California Peninsula.

1926–1988: No *Polioptila californica*

Joseph Grinnell was basically the father of California ornithology. As the founder and director of the Museum of Vertebrate Zoology at the University of California, Berkeley, he spent thirty years studying and cataloging the bird species of the state, and his decisions held great sway. Along with Alden Miller, Grinnell compiled all existing knowledge about California birds into a six-hundred-plus-page tome, heralded for decades as the foundation of ornithology in California.[12]

In Joseph Grinnell's 1898 monograph "Birds of the Pacific Slope of Los Angeles County," he provided a brief description of *Polioptila californica,* calling it the "black-tailed gnatcatcher"—illustrating why scientists use the Latinized names to keep things straight over time. He also detailed the species in 1904, when he saw one near

Palm Springs—40 kilometers east of where California gnatcatchers have ever been observed—and promptly shot it: "*Polioptila californica.* Black-tailed gnatcatcher. I secured [killed] a lone specimen, a female, on January first, two miles east of Palm Springs. I heard and recognized its call, and singled it out from among a scattered band of the plumbeous [*P. plumbea* at the time, now *P. melanura*]. The black-tail was being set upon and vindictively harried by a pair of plumbeous, which very plainly indeed resented its intrusion upon their domain. This bird was doubtless a straggler from the direction of Banning."[13]

But in 1926, Grinnell demoted the California gnatcatcher to a subspecies, arguing that all of the *P. californica* museum specimens actually belonged to a subspecies of *P. melanura,* the black-tailed gnatcatcher. He noted that there were many forms within the black-tailed gnatcatcher group but that the appearance of intermediate forms and the overlap in plumage characteristics suggested that these were all subspecies of one highly variable black-tailed gnatcatcher species. He justified combining everything under *P. melanura* based on its biogeographic similarities in distribution to another semiarid songbird, a type of sparrow called a brown towhee (*Pipilo fuscus* at the time). Notably, he described an obvious break in forms in both towhees and gnatcatchers at 30°N latitude on the Baja California Peninsula, indicating a California black-tailed gnatcatcher subspecies (*Polioptila melanura californica*) as distinguished from subspecies found farther south. Grinnell continued to treat the California gnatcatcher as a subspecies of black-tailed gnatcatcher throughout the remaining decades of his work on California birds.[14]

1989 to Present: California Gnatcatcher
Is a Species Again

In 1986, the ornithologist Jonathan Atwood completed his doctoral dissertation at the University of California, Los Angeles, titled

"The Gnatcatchers: *Polioptila californica, P. melanura,* and *P. nigriceps:* Species Limits, Vocalizations, and Variation." As a graduate student, Atwood reviewed all the available literature and more than two hundred museum specimens of gnatcatchers. He determined that, just as Brewster had thought a century earlier, the California gnatcatcher is indeed a distinct species from other gnatcatchers. In his assessment, Atwood highlighted the differences in plumage and calls that could be used in the field to distinguish California gnatcatchers, made possible by the transition from guns to binoculars as the go-to field equipment among birders. His work demonstrated that California gnatcatchers were selectively mating with each other and not with black-tailed gnatcatchers (*P. melanura*), a strong indicator of species-level distinctiveness.[15]

Based on Atwood's work, the California gnatcatcher regained official species status in the American Ornithological Union's checklist in 1989, primarily due to the overwhelming weight of morphological and biogeographical evidence. The species' distribution encompasses a thin band of coastal Southern California from Ventura County (about 34°N latitude) south to the southern tip of the Baja California Peninsula in Mexico. Its easternmost limit in the United States lies in central Riverside County, where it bumps up against the western limits of the black-tailed gnatcatcher (*P. melanura*).[16]

Subsequent genetic analysis by Robert Zink and Rachelle Blackwell supported the classification of California gnatcatchers as a distinct species. Using mitochondrial DNA, they determined that *P. californica* was genetically distinct from other arid-adapted gnatcatchers, including the black-capped gnatcatcher (*P. nigriceps*) to the south (mainly in Mexico) and the black-tailed gnatcatcher (*P. melanura*) to the east. From this analysis, they also determined that these three arid-adapted gnatcatcher species diverged quite recently, along with the white-lored gnatcatcher (*P. albiloris*) in Middle America. This genetic analysis agreed with the plumage differentiation by Brewster, the biogeography work of John Hubbard in 1974, and the morphology and behavior analysis of Atwood, all of which supported four species of gnatcatchers in the black-tailed group: *P. californica, P. nigriceps, P. melanura,* and *P. albiloris.* Since the late 1980s to today, the California gnatcatcher (*Polioptila californica*) has been officially recognized as a species by an overwhelming consensus of genetic, morphological, and behavioral data.[17]

The Battle over California Gnatcatcher Subspecies

Section 3 of the ESA defines a listable unit as a species, a subspecies, or a distinct population segment. In practice, the USFWS refers to all listable units as species, as defined by "any subspecies of fish or wildlife or plants, and any distinct population segment of any species of vertebrate fish or wildlife which interbreeds when mature." While the term *distinct population segment* was added to the ESA in a 1978 amendment, it was not defined until the USFWS and National Marine Fisheries Service (collectively the Services) issued a joint decision describing the term in 1996. A distinct population segment is the smallest taxonomic unit that can be listed under the ESA that is both distinct from other populations and significant to the maintenance of the species' genetic diversity or conservation. In 2003, the USFWS briefly considered changing the coastal California gnatcatcher's listing from a subspecies to a distinct population segment but decided against it. From here on, I will only discuss the arguments over the gnatcatcher's subspecies designation.[18]

Originally, Jon Atwood believed that the California gnatcatcher was a species with no distinct subspecies. But after several researchers critiqued his data analysis, Atwood reexamined his data and determined that three subspecies were evident: a northern coastal California gnatcatcher, *Polioptila californica californica,* from Ventura County down to 30°N latitude in the Baja California Peninsula, and two other subspecies farther south on the Baja California Peninsula (*P. c. margaritae* and *P. c. abbreviata;* fig. 4.1). There is broad agreement in the scientific literature on this 30°N southern boundary for the coastal California gnatcatcher, but the range boundaries for the two southern subspecies are not as clear, muting the consensus on their distinctiveness. On the Baja California Peninsula, 30°N latitude is a common biogeographic boundary for many taxonomic groups, possibly due to a seaway that existed there in the mid-Pleistocene epoch, which acted as a barrier for terrestrial species. There is also a significant change in coastal sage scrub vegetation south of 30°N where the sage scrub has more summer drought–adapted species, potentially reinforcing the California gnatcatcher's subspecies boundary through ecological or behavioral mechanisms. Finally, the preference that coastal California gnatcatchers display for elevations below 500 meters is not evident among the southern subspecies on the Baja California Peninsula, further suggesting that the *P. c. californica* is a distinct subspecies based on habitat preference.[19]

Atwood was not the first to identify distinguishable subspecies of California gnatcatchers. In the past, multiple subspecies have been suggested for the Baja California Peninsula, including *P. c. margaritae, P. c. nelsoni* (later named *P. c. pontillis*), *P. c. atwoodi,* and *P. c. abbreviata.* Much of the uncertainty in number and boundaries was due to the reliance on plumage color from museum specimens to distinguish them. Although even slight plumage color differences can signal divergence at the species level, these differences can be difficult to detect if the plumage color of museum specimens changes over time. Birds with gray feathers can exhibit foxing over time, when gray feathers become reddish or brown, and all feathers can appear darker owing to soot-staining from early twentieth-century air pollution. Determining whether either of these discolorations affect California

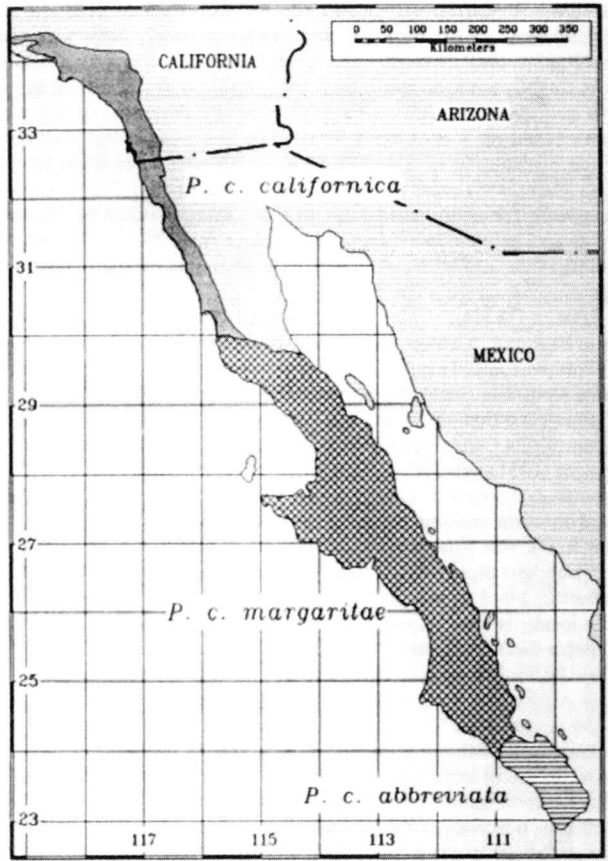

Figure 4.1. Distribution map of the three California gnatcatcher subspecies: P. c. californica *in solid gray,* P. c. margaritae *in hatched, and* P. c. abbreviata *in horizontal stripes.*
(Reproduced from Jonathan L. Atwood, "Subspecies Limits and Geographic Patterns of Morphological Variation in California Gnatcatchers [*Polioptila californica*]," *Bulletin of Southern California Academy of Science* 90 (1991): 118–133. OxyScholar © 1991.)

gnatcatcher museum specimens is complicated by the timing of these collections, as I explain later in this chapter.[20]

Although ornithologists have discussed the possibility of California gnatcatcher subspecies for almost a hundred years, the issue is a scientific and legal quagmire. If the coastal California gnatcatcher is not distinctly different from California gnatcatchers on the Baja California Peninsula south of 30°N, the California gnatcatcher could lose its protection under the ESA because, as a single species, it may not be rare enough to be listed. While in theory the USFWS can retain the coastal California gnatcatcher's protection regardless of its taxonomic situation, in practice, without evidence of its distinctiveness, it would be a more difficult listing to justify against the inevitable court cases and petitions to delist it. Lawsuits and petitions have already been filed against the gnatcatcher's listing as a subspecies. Given the ongoing threats that California gnatcatchers face on the Baja California Peninsula and our lack of population data across the gnatcatcher's entire range, delisting the coastal California gnatcatcher on the assumption that the species, as a whole, is safe from extinction is not a wise gamble.[21]

Soon after Atwood's published recognition of the three subspecies, Atwood, the Manomet Bird Observatory (where Atwood worked), and the Natural Resources Defense Council petitioned the USFWS to list the northern subspecies under the ESA. These petitions based their arguments for listing primarily on the extent of loss of coastal sage scrub, since no rangewide population survey data were available. In response, the USFWS issued a proposed ruling to list the coastal California gnatcatcher as endangered in 1991. In 1993, after two years of public comments and reviews, the USFWS issued its final rule, listing the coastal California gnatcatcher as threatened, rather than endangered, to assist California's 1991 Natural Community Conservation Planning policy (discussed in chapter 5). The

USFWS agreed that the coastal California gnatcatcher was a distinct subspecies, based on morphological and ecological characteristics, and was thus a listable unit.[22]

The U.S.-Mexico border splits the coastal California gnatcatcher's population roughly in half between the two countries. During listing decisions, the Services can acknowledge the impact that international boundaries have on species conservation, particularly when a species, subspecies, or distinct population segment may be subject to very different threats and levels of protection. In the gnatcatcher's case, the USFWS argued that the listing was necessary because the subspecies needed protection at least somewhere in its range: "Sufficient threats to the continued existence of the coastal California gnatcatcher exist in Mexico to warrant the listing of this subspecies throughout its range in Baja California. The government of Mexico has formally endorsed this conclusion and supports this listing action." By 2000, the Mexican government had still made no official moves to protect the California gnatcatcher, and the USFWS noted that "the already discontinuous gnatcatcher populations in the region may therefore be particularly susceptible to increased isolation and fragmentation due to ongoing unregulated habitat destruction."[23]

The listing decision was the start of decades of petitions and lawsuits seeking to contradict or question the science used to support the distinctiveness of the California gnatcatcher subspecies. The basis of the first listing lawsuit, brought in 1994 by the Building Industry Association of Southern California and other plaintiffs, made a procedural argument, stating that Atwood's raw data supporting subspecies of California gnatcatchers weren't made public, as required by law. Atwood's paper had undergone peer review and the paper was available to the public, and this met the USFWS's criteria for publicly available evidence. This wasn't just a picky procedural spat. The plaintiffs had learned of testimony that Atwood provided to the USFWS, in which he admitted to disposing of some of his raw data before publication. Furthermore, Atwood used a series of data manipulation methods—such as extrapolating missing information from averaged data—for analyzing gnatcatcher measurements among subspecies. While these practices are common and justifiable, they must still be explained in scientific publications in enough detail that the practices can be replicated. Unfortunately, Atwood offered no explanation for when he used these practices and why. Ultimately, after a lower court sided with the plaintiffs regarding their procedural arguments and temporarily delayed the coastal California gnatcatcher's listing, the data were released, and the plaintiffs dropped their suit.[24]

Until 2000, all the evidence supporting the distinctness of California gnatcatcher subspecies was based on morphology, ecology, and biogeography. As genetic sequencing technology became faster, cheaper, and easier, genetic data began to play a bigger role in taxonomy generally and in the California gnatcatcher's case specifically. Genetic technology has revolutionized evolutionary science and upended assumed knowledge in a wide variety of fields, including human evolution and biogeography. In the avian world, the number of cryptic species (multiple species that look identical but are genetically distinct) and false species (a single genetically identifiable species that looks like many) has been revealed to be far larger than most ornithologists anticipated. These discoveries reignited a long-simmering debate over whether the subspecies category is appropriate for birds at all.[25]

While the genetic work by Robert Zink and Rachelle Blackwell firmly supported the California gnatcatcher as a distinct species, genetic distinctiveness at the level of subspecies was another matter. Zink and his colleagues strongly argued against the designation of subspecies in the California gnatcatcher in a series of papers starting in 1998. Their research used genetic material collected from California gnatcatchers throughout the species' range in the United States and Mexico. As the technology improved, Zink's laboratory was able to test more individual birds across larger areas of their genome. Thus, while in 1998 his analysis was limited to the mitochondrial DNA (mtDNA) from just two gnatcatchers, by 2000 his laboratory was able to examine genetic material from sixty-four birds across the species' range.[26]

In these papers, one of which Jonathan Atwood helped to write, Zink and his coauthors concluded that California gnatcatchers did not display enough genetic distinctiveness to warrant subspecies designations. They used a threshold of least 1 percent difference in genetic characteristics called haplotypes, which are sets of genes or areas of the genome that tend to stick together and thus get inherited as a group over multiple generations. They also observed that individuals at the far northern end of the range, in northern Los Angeles County and Ventura County, had lower variability in their mtDNA. The authors interpreted this as evidence that the species was expanding its range northward, suggesting that the subspecies had not established in the area and differentiated sufficiently over time.[27]

In 2001, Zink and coauthors compared the genetic patterns of California gnatcatchers to five other native bird species in the region: black-tailed gnatcatcher, curve-billed thrasher (*Toxostoma curvirostre*), canyon towhee (*Melozone fusca*), cactus wren (*Campylorhynchus brunneicapillus*), and verdin (*Auriparus flaviceps*). They found that the cactus wren and verdin exhibited a much clearer genetic break at the 28°N to 30°N latitude boundary, reflecting that ancient midpeninsula seaway. Unlike the other species examined, the California gnatcatcher's genetic profile suggested that it had expanded northward from the Baja California Peninsula into the United States more recently, lowering the probability that it had been isolated long enough to differentiate into subspecies.[28]

Using a different approach, John Skalski and coauthors reanalyzed the available morphological data for California gnatcatchers with more sophisticated statistics and, like Zink, also determined that the coastal California gnatcatcher was not distinct. The authors cast doubt on the use of plumage colors to differentiate subspecies, especially using museum specimens when plumage color changes over time. The oldest gnatcatcher specimens were collected from Southern California, whereas newer specimens were more commonly collected from Mexico as naturalists explored southward into Baja California. But their conclusions were criticized by other scientists because they didn't cite or discuss earlier work that had controlled for aging plumage in museum specimens and still found plumage differences among subspecies. In its reconsideration of the gnatcatcher's listing status, the USFWS ultimately discounted the Skalski study because the study only used Atwood's original morphological data in its statistical analyses.[29]

Robert Zink is well known for criticizing the utility of subspecies designations for birds. Indeed, the interpretation of the phylogenetic species concept that Zink follows completely precludes the existence of subspecies. One critique of the Zink and Skalski work stated: "Zink's views on subspecies are not widely accepted by his peers … and reflect a rather extreme adherence to the phylogenetic species concept." But there is nothing inherently suspicious about Zink's attitude toward subspecies; scientists can have different opinions. His strongly stated views against subspecies in birds is likely what attracted research funding from development interests, which benefited from his science refuting distinct subspecies to support their lawsuits. Although this convenient synergy is common in the relationship between science and law—interest groups often fund science that they believe will support their positions—these funding relationships tend to cast doubt on the objectivity of the science. Doubts are strengthened when there is a lack of transparency for funding sources, as was the case for a couple of papers that Zink published. Of course, the building industry and others argue that environmental groups likewise seek out scientists who are predisposed to find or support listable units—one biologist referred to clusters of scientists with this philosophical leaning as "species cartels." Atwood's research was funded by environmental organizations and local land conservancy groups, thus fueling the insinuation that Atwood's work was biased, too.[30]

As Zink's and Skalski's evidence against a distinct subspecies began to accumulate, the Pacific Legal Foundation, representing agricultural and development interests, petitioned the USFWS to delist the gnatcatcher in 2011. Again, after a lengthy public comments period, the USFWS determined that the overall weight of evidence still supported the subspecies designation and rejected the petition.[31]

In its 2011 decision to uphold the listing of the coastal California gnatcatcher, the USFWS specifically requested research using nuclear DNA to further resolve the subspecies conundrum. In response, Zink used mtDNA and nuclear DNA in a subsequent genetic study of the California gnatcatcher, and he again argued in a 2013 paper that the northern populations were not a distinct subspecies from the populations on the Baja California Peninsula. In this new study, he and his coauthors examined a larger number of nuclear DNA genes as well as mtDNA. They focused on eight specific areas of the chromosomes to determine whether there were discrete groupings within California gnatcatcher subspecies. If these groupings were observed, the groupings would likely reflect genetic differences among subspecies.[32]

Zink's new genetic work received considerable criticism in a 2015 paper by John McCormack and James Maley, highlighting that Zink's methodological choices were contrary to what he and the USFWS had recommended in the past. Zink wrote a rebuttal to McCormack and Maley in 2016—not in time for a six-member scientific advisory panel to consider it when they supported the subspecies designation, but in time for the USFWS biologists to respond to it in their final decision to maintain the subspecies' listing.[33]

Genetic techniques are not the only new approach that can be applied to taxonomic and conservation questions. A methodology called *ecological niche modeling* (also called *species distribution models*) can be used to determine ecological distinctiveness as described in the ESA. An ecological niche is a collection of environmental factors required for individuals of a species to survive, such as minimum and maximum temperatures, average precipitation, vegetation, soil type, and elevation. Ecological niche analysis uses presence/absence or abundance data paired with environmental data to determine an organism's niche. Ecological niche models can be used to compare niches between species or subspecies, in order to determine whether the niches are significantly different—if so, the difference lends support to taxonomic distinctiveness.

In that same 2013 genetics paper described above, Zink and co-authors combined the genetics results with an ecological niche model. Their ecological niche model agreed with their analysis of genetic data; the northern coastal subspecies was a recent range expansion from the original populations on the Baja California Peninsula, with no clear breaks or distinctions in its distributional range to signal discrete subspecies. Their ecological niche model was also criticized, in this case for using only temperature and precipitation data and ignoring other factors that previous research had suggested were important for the gnatcatcher and other coastal sage scrub species. At the time, it was not yet clear how sensitive ecological niche models are to included and excluded variables. Curiously, when others reanalyzed Zink and colleagues' niche modeling data, they found clear evidence of niche differentiation among subspecies, especially at the 28°N to 30°N latitude boundary.[34]

Evidence of niche differentiation among California gnatcatchers at this 30°N latitude suggests two possible explanations: (1) the gnatcatchers above that latitude choose different habitat than those below it, or (2) the gnatcatchers above that latitude have different habitat available to them but might use the southern-style habitat if it were available. The first option would be evidence for distinct subspecies; the second would not. Zink and his coauthors chose option 2, and scientists supporting distinct subspecies chose option 1. The methods that can determine which explanation is correct require substantial data collection throughout the California gnatcatcher's range. These data do not yet exist.

The publication of Zink's 2013 paper triggered another petition by the building industry and others to delist the coastal California gnatcatcher. Had the petition been successful, it could have released almost 81,000 hectares of protected area to development. Criticisms of Zink's work, comments received from the public, and the reviews of a six-member scientific advisory panel—convened to provide advice to the USFWS—all factored heavily in the USFWS's 2016 decision to uphold the listing of the coastal California gnatcatcher as a distinguishable subspecies.[35]

Aside from the specific criticisms regarding Zink's work, the size and fervor of the debate reflects a broader commentary on our love affair with technology and new things. Genetic sequencing technology is advancing quickly and becoming less expensive, allowing for the production of more genetic data and even the sequencing of a species' entire genome. But sequencing a genome does not tell you what genes are doing and whether they have large or small influences over an organism's behavior, appearance, or mating preferences. We need to be cautious about elevating genetics data over other kinds of data to determine taxonomy and delineate species. A century of morphological, ecological, and behavioral data supports the distinctiveness of the northern California gnatcatcher subspecies, and therefore it should take equally broad and persuasive evidence to overrule those studies.[36]

The USFWS continues to request more research. The agency has specifically asked for a genetic analysis that includes more of the California gnatcatcher's genome, using a much larger sample of birds throughout the species' range in the United States and Mexico. These tests may still fail to find parts of the genome that accurately reflect the state of divergence of the subspecies; it is a bit like finding a needle in a haystack. If more of the genome is sampled without finding significant differences, the results would lend support to Zink's position that the California gnatcatcher does not have distinct subspecies. If habitat loss continues along the U.S.-Mexico border—or if a 10-meter-high wall appears—the populations on the northern side of the border will likely genetically diverge from those on the southern side over time anyway, owing to the loss of reproductive connectivity of California gnatcatchers across the border.[37]

At this time, I don't honestly know if the coastal California gnatcatcher is distinctive at the level of a subspecies or distinct population segment. I have never observed California gnatcatchers south of the U.S. border, and I haven't seen the museum specimens or the genomic data. I am in no position to judge. I am fairly confident, though, that had the coastal California gnatcatcher not been listed under the ESA, the loss of coastal sage scrub would have continued at a faster pace than has occurred over the past thirty years. During my sabbatical in 2016–2017, I was stunned to see how far housing tracts had pushed up into the foothills and out into the valleys, where agricultural fields and scraps of coastal sage scrub had existed in the early 1990s. I also suspect that the California gnatcatcher's listing prevented other coastal sage scrub species from ending up on the ESA list. As I'll explain in chapter 6, species listed under the ESA incentivize landowners to participate in regional-scale habitat conservation, benefiting listed and nonlisted species alike. Whether the coastal California gnatcatcher continues to provide this incentive in Southern California is an open question.

5

The Gnatcatcher and the ESA

America's land once seemed inexhaustible. There was always more
of it beyond the horizon. Until the twentieth century we displayed
a carelessness about our land, born of our youthful innocence and
desire to expand. But our land is no longer an open frontier. Americans not only need, but also very much want to preserve diverse
and beautiful landscapes, to maintain essential farm lands, to save
wetlands and wildlife habitats, to keep open recreational space near
crowded population centers, and to protect our shorelines and
beaches. Our goal is to harmonize development with environmental quality and to add creatively to the beauty and long-term worth
of land already being used.

President Richard Nixon,
State of the Union Address, February 15, 1973

In March 1993, the coastal California gnatcatcher received protection as a threatened species under the federal Endangered Species Act
of 1973, one of the most powerful and controversial environmental
laws in the world. The ESA is the legal "stick" that is essential to the
success of the Natural Community Conservation Planning policy that
aims to boost regional habitat conservation in California (discussed in
chapters 6 and 7). Before the coastal California gnatcatcher's listing,
few landowners were eager to enter into NCCP conservation agreements. After its listing, NCCP plans multiplied quickly.[1]

The coastal California gnatcatcher's protection under the ESA is itself a dramatic story, amplified by multiple lawsuits. These lawsuits affect ESA implementation around the country, entwined in a dense legal tapestry that defines the ESA's potential and its limitations for protecting our biological heritage.

A Brief History of Biodiversity Conservation

Protecting species from extinction was not invented by the United States in 1973. Throughout human history, taboos and cultural prohibitions on hunting, fishing, and collecting have ensured the long-term availability of vital resources. For example, indigenous tribes in North America held strict taboos against harvesting adult passenger pigeons at roosting sites. They waited for the end of the nesting cycle to harvest only the preflight squabs, so that fledged juveniles and breeding adults would produce more chicks the next year. These harvesting taboos seemed ridiculous to white settlers, who were astonished at the flocks of hundreds of millions of pigeons that would crush entire forests under their weight. In hindsight, these taboos were a smart management practice: the species was overharvested to extinction after just fifty years of hunting by white settlers.[2]

Public support for harvesting prohibitions to protect declining species grew with the loss of charismatic species, such as the passenger pigeon. The first U.S. law to regulate species use was the Lacey Act in 1900, prohibiting the trade of game species across state lines. The Lacey Act was motivated by the slaughter of wading birds for their feathers, used to decorate women's hats. During his years as president from 1901 to 1909, Theodore Roosevelt protected over 93 million hectares of bird habitat by establishing national refuges and reserves, mostly by executive order. Later, the 1918 Migratory Bird Treaty Act protected all birds on both their breeding and nonbreeding ranges, ensuring that efforts to protect birds in one country were not negated by mass killings in other countries. The Migratory Bird Treaty Act was the first environmental law to use the word "take" in a conservation context, referring to the direct killing, harassment, or transport of live or dead individuals, including eggs. In the ESA, the definition of *take* was expanded to include the destruction of occupied habitat and to clarify the actions that qualify as harassment.[3]

Later, the Fish and Wildlife Coordination Act of 1934 conserved game species and their habitats, and the Bald Eagle Protection Act of 1940 prohibited take or possession of bald eagles (*Haliaeetus leucocephalus*) or their parts; the act was amended in 1978 to include golden eagles (*Aquila chrysaetos*). As time passed, endangered species laws incrementally increased the level of protection for animals and plants: the Endangered Species Preservation Act of 1966 protected vertebrate animals in national wildlife refuges; the 1969 Endangered Species Conservation Act protected threatened species brought into the United States from elsewhere; and the Wild Free-Roaming Horses and Burros Act of 1971 and the 1972 Marine Mammals Protection Act protected wild horses and marine mammals, respectively. Finally, the ESA as we know it today was passed by the U.S. Congress and signed by President Richard Nixon in 1973.[4]

The Convention on International Trade in Endangered Species of Wild Fauna and Flora (CITES) comes close to matching the ESA's powerful bite. An international treaty ratified by most countries, CITES was finalized and came into force in 1975. The United States is a signatory country and is therefore bound to enforce it. CITES regulates the international trade in plants and animals that are on the International Union for Conservation of Nature's "Red List"—species considered threatened with extinction. National signatories are obligated to assign hefty fines and jail terms for importing Red List organisms or their parts. In the United States, CITES and the ESA work in tandem to protect listed species moved across U.S. borders.

Protecting individual plant and animal species is only half of the conservation challenge. Organisms need a place to live, too. Safeguarding habitat and ecosystems stretches back into premodern times, exemplified by the hunting reserves owned by European royals and sacred groves managed by community elders to honor spirits and valued species. Yellowstone National Park, the first publicly owned protected area in the world, was established in 1872 in the United States to preserve unique scenic qualities and the rapidly disappearing American bison (*Bison bison*). The National Wilderness Preservation Act of 1964 and the 1966 Endangered Species Preservation Act designated wilderness areas across the nation and codified the national wildlife refuge system, respectively.[5]

The Endangered Species Act of 1973

The Endangered Species Act of 1973 is the cornerstone biodiversity conservation policy in the United States. The ESA promotes both the survival *and* the recovery of species at risk of extinction, prioritized over economic considerations. Championed in the U.S. House by Democrat John Dingell of Michigan and in the U.S. Senate by Republican Mark Hatfield of Oregon and Democrat Harrison Williams of New Jersey, it passed almost unanimously in both houses and was signed by President Nixon. The Endangered Species Act is enforced by the U.S. Fish and Wildlife Service for terrestrial and aquatic species and the National Marine Fisheries Service for marine species.[6]

Broadly speaking, the ESA works in four ways: banning international trade in listed species by enforcing CITES; discouraging the federal government from conducting activities that jeopardize listed species; prohibiting private individuals and organizations from killing or harming listed species without a permit; and requiring the Services to develop species recovery plans. As of 2019, 1,662 species were listed as threatened or endangered, and 41 had recovered and were delisted — 27 of them in the past ten years. Although critics like to point out that such numbers represent a recovery rate of only 2 percent, other statistics are more encouraging. So far, the ESA has preserved almost 300 species that would have gone extinct without its protection and has prevented countless others from entering an extinction spiral. Most telling, unlisted species are far more likely to go extinct than species that enjoy the ESA's protection. In 2019, the 71 U.S. species that were presumed extinct had been lost before they were listed. Ultimately, the goal of the ESA is to make itself redundant — to protect enough individuals and habitat, and mitigate enough threats, so that all species can be delisted due to their recovery.[7]

Well-known success stories include the bald eagle, peregrine falcon (*Falco peregrinus*), grizzly bear (*Ursus arctos horribilis*), California condor (*Gymnogyps californianus*), American alligator (*Alligator mississippiensis*), and gray wolf (*Canis lupus*). The peregrine falcon and bald eagle recovered completely and were delisted in 1999 and 2007, respectively. The Yellowstone population of grizzly bears was delisted in 2017, but other populations remain listed elsewhere. Although the American alligator has officially recovered, the species remains listed as "threatened" to protect its doppelgänger, the threatened American crocodile (*Crocodylus acutus*), from being killed due to mistaken identity. The California condor is still listed as endangered, but a decades-long captive breeding program at the San Diego Zoo has reestablished wild populations in the American Southwest. The listing status of the gray wolf is complicated because it is protected and managed by regional distinct population segments, a controversial taxonomic category with hazy scientific support (see chapter 4).[8]

The 1982 congressional amendment was one of the most significant for the implementation of the ESA. The 1982 amendment directed the Services to consider economic impacts in two of their three major decisions: identification of critical habitat and options included in recovery plans. In this amendment, Congress explicitly emphasized that listing decisions should not consider economic impacts because extinction risk is based purely on biological circumstances. The 1982 amendment also allowed for some negative impacts to listed species by creating "incidental take permits" in Section 10 of the act. Often called "10(a) permits," these allow permit holders to kill, harm, or harass listed species but are granted only in conjunction with an approved Habitat Conservation Plan (HCP). These HCPs form the basis of conservation efforts for many listed species on private property if private landowners seek 10(a) permits—as is the case for the California gnatcatcher.[9]

Public resistance to the ESA has grown as it has extended its reach onto private property. By the 2000s, two-thirds of listed species could be found on privately owned land, and one-third of listed species could be found only on private property. At the time of the coastal California gnatcatcher's listing, about half of its remaining habitat was privately owned. ESA enforcement can have substantial economic ramifications for landowners, although the costs of species listings are far more obvious than their benefits. Private property advocates argue that the loss of the economic value of private property due to land use restrictions is a violation of the Fifth Amendment of the U.S. Constitution, which prohibits the federal government from seizing private property without adequate compensation—commonly known as the Takings Clause. The relevance of the Fifth Amendment is clear in cases of eminent domain when the government acquires land outright, but less so when the government enforces regulations that restrict only some land-use activities, as the ESA often does. Private property owners have had little success when using the Takings

Clause argument against the ESA in court. The courts have repeatedly decided that species extinctions are a public harm, and thus land-use restrictions fall under the police power of state and do not require compensation.[10]

The ESA has survived numerous attempts in Congress to weaken it. The Center for Biological Diversity tallied dozens of legislative attempts by the 115th Congress alone to weaken the ESA or repeal it completely. Two proposals would have devolved enforcement power to the states, one requiring the consent of every governor of every state in which a species was found before it could be listed and the other prohibiting species found in only one state to be listed under the ESA—if made retroactive, the latter proposal would have delisted over a thousand species. A different proposal exempted fossil fuel extraction activities from ESA restrictions, and another proposal would force the Services to factor economic impacts into the listing decision, in addition to the critical habitat and recovery decisions.[11]

In a 2017 interview with *National Geographic,* journalist Christopher Ketcham asked then-retired Representative John Dingell if he would have been able to get the ESA passed in the current Congress. Dingell responded, "I don't think I could pass the Lord's Prayer in that nuthouse."[12]

For the sake of brevity, I discuss only the ESA sections that are the most relevant to the conservation efforts for the California gnatcatcher: Sections 4, 7, 9, and 10. How these sections are written, interpreted by the courts, and enforced by the USFWS drives much of the California gnatcatcher's story and the lessons learned from its protection.[13]

Section 4 and 4(d) Rules

Section 4 describes the bureaucratic decision-making processes for listing species, designating their critical habitat, and creating species recovery plans—the meat and potatoes of the ESA. Lawsuits and petitions have challenged the listing and critical habitat decisions for the coastal California gnatcatcher, but none has involved the recovery plan process because, as of 2020, the California gnatcatcher had no official recovery plan.[14]

The first decision that the Services must make is whether a plant or animal is a listable unit. This listable decision involves two criteria, both of which the species should meet: taxonomic distinctiveness (see chapter 4 for the gnatcatcher) and rare and declining population trends (see chapter 3 for the gnatcatcher). I say "should meet" because, although it is technically possible for the Services to list almost anything, they are guaranteed to be swamped by lawsuits and petitions if their listing decision seems arbitrary or contrary to science. Indeed, the growing corpus of lawsuits over listing decisions influences the listing process. Petitions and lawsuits to delist the coastal California gnatcatcher have been centered on both taxonomy and rarity points: there are no distinct subspecies, and there are large populations in Mexico. So far, all of the lawsuits and petitions against the coastal California gnatcatcher's listing have failed.[15]

Scientific ambiguity and lack of biological data are common themes in the listing process. Ambiguity and uncertainty delay listing decisions and create challenges for crafting and implementing sound policy, which often assumes high levels of certainty. Although the Services must rely on the "best available scientific and commercial data" to make their decisions, the data are not always available, sufficient, or unanimous. Scientists often reserve judgment when data are equivocal and may change their opinions with new data (see chapter 4). This is a normal part of the scientific process. When scientific evidence for listing is equivocal, the Services' biologists use their discretion to determine whether a listing decision is supported by the weight of the evidence. This discretionary power is not unique to the Services. Recognizing that subject matter expertise lies with the executive branch, Congress often writes laws that are vague to provide some latitude and flexibility to implementing agencies. The U.S. Supreme Court upheld agency discretion when enforcing laws and regulations in a landmark case, *Chevron U.S.A., Inc. v. Natural Resources Defense Council, Inc.*, which established this discretion in what is known as the Chevron deference doctrine. In the case of the California gnatcatcher, the USFWS has consistently decided that the weight of morphological and ecological data support a listable subspecies of coastal California gnatcatcher.[16]

If the Services find that a species is listable, they then must determine if the species meets the criteria for being threatened or endangered. The term *threatened* is defined as "likely to become an endangered species within the foreseeable future throughout all or a significant portion of its range" and *endangered* as "in danger of extinction throughout all or a significant part of its range." The Services must also identify the factors that are responsible for a species' imperiled status, again using the best scientific and commercial data available.[17]

Similar to listing decisions, delisting and downlisting from endangered to threatened must be based solely on scientific data and not economic impacts. Species are delisted or downlisted when the species no longer meets the listing criteria: when it is no longer threatened or endangered or when its taxonomic distinctiveness is called into question based on new data. The Services can separate a species into several distinct population segments and uplist, downlist, or delist any of them separately. The grizzly bear and gray wolf species are listed and managed as distinct populations, for example. In unfortunate cases, species are delisted because they have gone extinct.

Section 4(d) of the ESA allows the Services to provide threatened species with additional protection or unique management actions if warranted; these are called 4(d) special rules. These special rules are necessary because, whereas endangered species receive all of the protections outlined in Section 9, threatened species do not. The 4(d) special rules are unique to species and circumstances and are meant to improve conservation outcomes in some way. When the USFWS decided to list the coastal California gnatcatcher in 1993 as threatened instead of endangered, the agency simultaneously issued a 4(d) special rule. This special rule purposefully positioned the coastal California gnatcatcher as the flagship species for the NCCP policy. The special rule linked incidental take permits to NCCP plans, capped the total amount of losses of gnatcatcher pairs and habitat area under incidental take permits to 5 percent while NCCP plans were in development, and delegated permit approval authority to the California Department of Fish and Wildlife (then called the California Department of Fish and Game). Section 4(d) rules are meant to provide flexibility for policy experimentation, such as the NCCP policy for the California gnatcatcher, and to provide more discretion to state governments. More generally, Section 4(d) was meant to make the ESA less of a political target during the reauthorization process.[18]

Until 2019, the USFWS automatically granted threatened species the same protections as endangered species under a Section 4(d) general regulation, referred to as the 4(d) blanket rule. The NMFS does not use a blanket rule but rather decides on a case-by-case basis whether to extend the Section 9 protections to threatened species. On August 27, 2019, the USFWS ended its blanket rule and adopted the NFMS's case-by-case system. Luckily, that rule change was not retroactive, otherwise the coastal California gnatcatcher and all other threatened species would have lost many of their legal protections.

The rule change was just one of many that were pushed after the 2016 U.S. election. The USFWS also decided on August 27, 2019, that when weighing the risk of extinction, its interpretation of foreseeable future would be notably shortened to a time frame with greater certainty. This policy change decreases the likelihood that species threatened solely by climate change and its uncertain impacts will receive ESA protection. Earlier in 2019, the USFWS had announced that it was revoking its compensatory mitigation policy that aimed for no net loss of habitat and a goal of net conservation gain. As you might expect, representatives of the Southern California development industry—and the law firms that represented them—expressed full support for these and other streamlining changes. Attorney Jonathan Wood with the Pacific Legal Foundation said, "By relaxing restrictions, landowners will have more reason to cooperate in the recovery of threatened species, and environmentalists will have less incentive to litigate these issues." As discussed in the next chapter, the experience of the NCCP policy does not support this level of optimism. The 2019 changes to ESA implementation practices did not benefit the public service careers of the two political appointees in the Department of Interior who pushed for them—both resigned from scandals before any changes were finalized.[19]

After listing, the Services must next delineate critical habitat, which is the area needed for the species' conservation regardless of whether the species currently occupies it. The courts have interpreted conservation to mean habitat that is important for both the survival of the species *and* its recovery, allowing for the designation of areas that are not currently occupied but were historically or are expected to be in the future. When designating critical habitat, the Services weigh the ecological benefits of the location and amount of habitat they designate with the negative economic impacts of doing so. The methods that the Services use to calculate the economic impact of critical habitat designations have also been the subject of many lawsuits, including two involving the coastal California gnatcatcher.[20]

Critical habitat designations primarily affect the activities of federal agencies under Section 7, requiring that these activities avoid modifying designated habitat in ways that generate harm to listed species. Designated critical habitat can include both public and private property. Within designated critical habitat areas, activities conducted by federal agencies or funded by federal agencies (a "federal nexus") that might destroy or alter habitat must receive a permit from the Services. Outside of designated areas, and in the absence of federal involvement, activities that damage or destroy a listed species' habitat are legal, as long as they do not constitute a take of protected species.[21]

In the years following the 1978 congressional amendment codifying critical habitat, the USFWS avoided designating critical habitat whenever possible. By 1997, the USFWS had designated critical habitat for only a handful of listed species. In 2000, just 10 percent of listed species had designated habitat, rising to 41 percent by 2009. Many of these habitat designations were forced on the USFWS by lawsuits. Why? Although habitat is important to species conservation, designating habitat is impossible if the USFWS doesn't have enough information—such as rangewide species surveys—to determine the location of the habitat. And, since the Services have to open their proposed critical habitat maps for public comment, the ninety-day comment period gives landowners a lot of advance warning if their property might land in a designated area. For this reason, the USFWS long argued that designating critical habitat is counterproductive if it incentivizes landowners to destroy habitat before the designation becomes official. Somewhat contradictorily, the USFWS has also argued that critical habitat designations are redundant with enforceable actions under other sections of the ESA and thus are unnecessary.[22]

In the gnatcatcher's case, the USFWS declared that designating critical habitat would not be prudent; it would place the species at further risk due to intentional habitat destruction on private property. The USFWS provided evidence of eleven cases where landowners had preemptively destroyed coastal sage scrub before critical habitat was designated. A legal chronology of the California gnatcatcher's designated critical habitat illustrates just how much chaos this one decision—or lack thereof—can generate for the Services. After the USFWS decided that it would not designate critical habitat for the gnatcatcher, the Natural Resources Defense Council and the Audubon Society sued the USFWS in 1997 to compel it to do so. After some delay, the USFWS designated a total of 207,890 hectares of coastal sage scrub across five counties (Los Angeles, San Bernardino, Riverside, Orange, and San Diego), 83 percent of which was privately owned. The USFWS then faced three lawsuits in 2000–2001: one from the NRDC over the exclusion of NCCP areas and two from development interests for underestimated economic impacts of the designation. In 2003, the USFWS published a revised critical habitat area, representing a reduction of 3.5 percent to 200,595 hectares across six counties (the previous five plus Ventura).[23]

Around this time, the USFWS publicly declared that the resources it was expending on critical habitat designations, and the associated lawsuits, were preventing it from doing anything else. This claim was undercut by a pair of U.S. General Accounting Office reports, which emphasized that most of the lawsuits that the USFWS faced over critical habitat were mainly for failing to designate it—including the NRDC's lawsuit over the coastal California gnatcatcher's habitat. Instead, the General Accounting Office suggested that the problems with the critical habitat process stemmed from a lack of guidance to ensure consistent and transparent decision-making.[24]

After an extensive public comment period, in 2007 the USFWS revised the California gnatcatcher's critical habitat area downward by 62 percent to only 79,846 hectares that had a clear federal nexus.

This area included some NCCP-enrolled lands but excluded areas enrolled in approved HCPs or Multiple-Species HCPs. Over 80 percent of the final designated critical habitat area was within private ownership. Of the thirteen critical habitat units, there was a notable reduction of Unit 10 from 80,913 hectares to 11,141 hectares in western San Bernardino and Riverside Counties, around the I-10/I-215 interchange. Unit 10 had one of the most expensive estimated economic impacts ($10,408/hectare), and the revised designation excluded the Santa Ana River basin, an area noted by the USFWS as a potentially critical connector of small gnatcatcher populations (see box 7.1). The 56 percent reduction in critical habitat in Unit 13 was also troubling, as that area of northern Los Angeles County and Ventura County can connect isolated northern gnatcatcher populations to larger populations farther south and allow the species to shift its range northward; it is thus an important area for the California gnatcatcher's climate adaptation (see chapter 3).[25]

Given the stress that critical habitat designations generate for private landowners, the USFWS tries to avoid designating private land with no expected federal nexus; this was a consideration for the California gnatcatcher. However, since a large proportion of listed species' habitat is privately owned, often the USFWS must include private property in critical habitat designations. Within a designated critical habitat area, landowners planning to modify or destroy habitat must get their land surveyed to assure that there are no listed species. If a listed species is found, landowners must develop an HCP for their land in order to receive a Section 10(a) incidental take permit (more about those below), or they can join an existing plan such as one under the NCCP policy (see chapter 6). If landowners don't go through the permit process and they destroy occupied habitat, they face the penalties outlined in ESA Section 11. As you might guess, this establishes an incentive to maintain one's land free of listed species and their habitat before these surveys—a scorched-earth approach once advocated by the National Association of Home Builders and clearly undertaken by at least eleven landowners in Southern California. Permit mechanisms in the NCCP policy were designed to disincentivize this habitat-destroying behavior.[26]

The ecological effectiveness of critical habitat designations is an open question. Some studies have found no effect of designations on the status of listed species or the rate of land use change, while others have found some evidence for improved outcomes with designations. Critical habitat designations may also boost property values within these areas, perhaps due to the anticipation of preserved open space, perversely increasing development pressure and habitat loss. As of 2018, the critical habitat designation for the California gnatcatcher had prevented development on nearly 40,500 hectares in the region, but it is difficult to say whether this has caused positive or negative changes in land use elsewhere. A 2007 economic impact assessment performed for the USFWS estimated that designating over 200,000 hectares as critical habitat for the coastal California gnatcatcher would result in negative economic impacts of almost $1 billion (adjusted 2002 dollars) between 2002 and 2025. As estimated by the Transportation Corridor Authority, that 200,000-hectare critical habitat area would be responsible for 2.2 to 5.6 percent fewer jobs in the region, 3.6 to 8.5 percent fewer housing units built, and costs to the regional economy of $0.4 to $5.5 billion over a twenty-year period. These economic estimates were completed before the USFWS reduced the California gnatcatcher's critical habitat by 62 percent and before the 2008 foreclosure crisis and subsequent recession depressed home buying, home prices, and construction for years.[27]

Species recovery to the point of delisting is the ESA's ultimate goal. Despite this, neither species recovery nor recovered species were defined by Congress in the ESA. Therefore, translating recovery goals into specific, quantitative targets is at the discretion of Services biologists. Recovery plans must review the species' biology and the threats it faces, specify recovery targets, and list action options and their costs. Recovery plans are either written in-house by Services personnel or contracted out to experts. The 1988 ESA amendment added a required ninety-day public comment period to the recovery plan process, similar to the listing and critical habitat designation stages. Given the impact that recovery plans have on many landowners and industries, recovery plans are usually drafted with significant input from a variety of public and private stakeholders.[28]

Recovery plans suffer from a variety of shortcomings. Recovery criteria are often vague and inconsistent, lack guidelines for monitoring, and sometimes set the target recovery population size *below* the population size when the plan was written. Some of these issues stem from the difficulty of defining "recovery"; there are disagreements within the scientific community as to what these recovery targets should represent. Some species are rare by nature and more vulnerable to extinction, even without human interference—population recovery targets for these species might never be achieved even under the best circumstances. That said, species with recovery plans are more likely to improve and be downlisted or delisted than species that do not have one. Species covered under multispecies recovery plans are less likely to improve, but a majority of species are covered by them.[29]

In theory, the USFWS uses a science-based system for prioritizing recovery funding, weighted by the degree of threat, potential for recovery, and the distinctiveness of the species. Unfortunately, the recovery stage is severely underfunded and profoundly biased. Annually, a handful of species receive millions of dollars for recovery projects, while many receive a few thousand dollars or nothing at all. Funding allocation is influenced by lawsuits, congressional earmarks for charismatic species—especially bald eagles and peregrine falcons—and the political clout represented within a species' range. Species in populous states such as California and Florida receive recovery funding disproportionate to their risk or distinctiveness, particularly if those states are represented by Democrats. The amount of money spent really matters; species that receive the most funding are most likely to recover.[30]

Recovery plans are not a highly prioritized part of the listing process. Plans merely provide guidance, not regulatory authority. Even though HCPs are expected to increase the likelihood of species recovery, these plans cannot substitute for a recovery plan because they focus on conservation—preventing a species from further decline. There is no deadline after listing that recovery plans have to be approved or completed, and no penalty if the actions outlined in approved recovery plans are never taken. Even so, the Services and other federal agencies can still be successfully sued over their decisions and actions involving recovery plans.[31]

The Services have made some progress on developing recovery plans. In 1994, 54 percent of listed species had plans, and by 2010 it was up to 86 percent. Over 90 percent of listed bird species now have recovery plans, spanning decades of recovery actions. But the coastal California gnatcatcher has been listed for almost thirty years and still does not have a recovery plan—the NCCP policy is designed to prevent the gnatcatcher's extinction, not actively promote its recovery, and it therefore does not set any recovery targets. The coastal California gnatcatcher had a recovery priority number of 9C in the most recent USFWS Recovery Report (FY2013–2014); at the time of its listing in 1993, it had a higher priority of 3C.[32]

Section 7—Federal Nexus

Section 7 may be familiar to readers who have heard of Section 7 consultations or federal nexus. Requiring federal agencies to institutionalize biodiversity conservation, Section 7 is one of the strongest sections in the ESA. This section requires all federal agencies to consult the Services when species might be placed at risk by such activities as building roads or dams or releasing pollutants or noise. When the Services provide ESA consultation to other federal agencies or to nongovernment organizations receiving federal money, it is called a Section 7 consultation. Section 7 consultations are triggered by the 1970 National Environmental Policy Act, a broad law that ensures that federal agencies will comply with federal environmental laws such as the ESA, the Clean Air Act, and the Clean Water Act. A Section 7 consultation is triggered by a federal nexus: land owned by the federal government, activities conducted with federal funding by nonfederal organizations, and activities conducted by federal agencies regardless of land ownership—including granting permits. A federal nexus occurs whenever the Services issue incidental take permits as part of an approved HCP or NCCP plan. Section 7 also delineates an Endangered Species Committee, known as the God Squad, which is a collection of federal agents that can exempt a species from protection under unique circumstances. Although Section 7 generates a lot of complaints, the consultation process rarely scuttles projects. Out of more than eighty-eight thousand Section 7 consultations conducted between 2008 and 2015, not a single project was halted or required extensive changes. Whether this is a sign of a well-functioning consultation process or a process that is lacking rigor is another matter.[33]

The USFWS performs Section 7 consultations involving the coastal California gnatcatcher for projects with federal funding, such as county transportation and sewer projects, and for military activities at the U.S. Marine Corps' Camp Pendleton and Miramar Air Station and the U.S. Navy's Seal Beach Detachment Fallbrook. As of 2016, the USFWS had completed 320 formal Section 7 consultations for the coastal California gnatcatcher, with none requiring modification of a proposed action to ensure compliance with the ESA. But the coastal California gnatcatcher was involved in a lawsuit regarding a Section 7 consultation outcome: the case of the USFWS's no jeopardy opinion that allowed the San Joaquin Hills toll road in Orange County to destroy vital gnatcatcher habitat, which I discuss in chapter 6.[34]

Military installations also fall under the 1960 Sikes Act (16 USC 670a–670f), which requires the Department of Defense to manage natural resources using sound conservation science through the use of Integrated Natural Resources Management Plans. The management plans for Camp Pendleton, Miramar, and Fallbrook include such activities as invasive species control and habitat restoration that benefit California gnatcatchers and a dozen other listed species, including the California least tern (*Sternula antillarum browni*), least Bell's vireo (*Vireo bellii pusillus*), and southwestern willow flycatcher (*Empidonax traillii extimus*). Camp Pendleton's importance to biodiversity conservation in the region cannot be understated. It is 50,590 hectares of natural habitats connecting coastal preserves in Orange County with those in San Diego County (fig. 5.1). As the largest landowner in the United States, the U.S. military is one of the most consequential property owners in the nation; military bases support some of the highest densities of ESA-listed species across all federal properties.[35]

Despite the negligible effect that consultations have on project plans, Section 7 decisions generate a large number of ESA lawsuits. In fact, one of the first ESA lawsuits in the 1970s involved Section 7 and a rare, finger-sized fish. Once referred to as "the two-inch terror" by former Senator Howard Baker from Tennessee, the federally endangered snail darter (*Percina tanasi*) held up the completion of the Tellico Dam on the Little Tennessee River, constructed by the

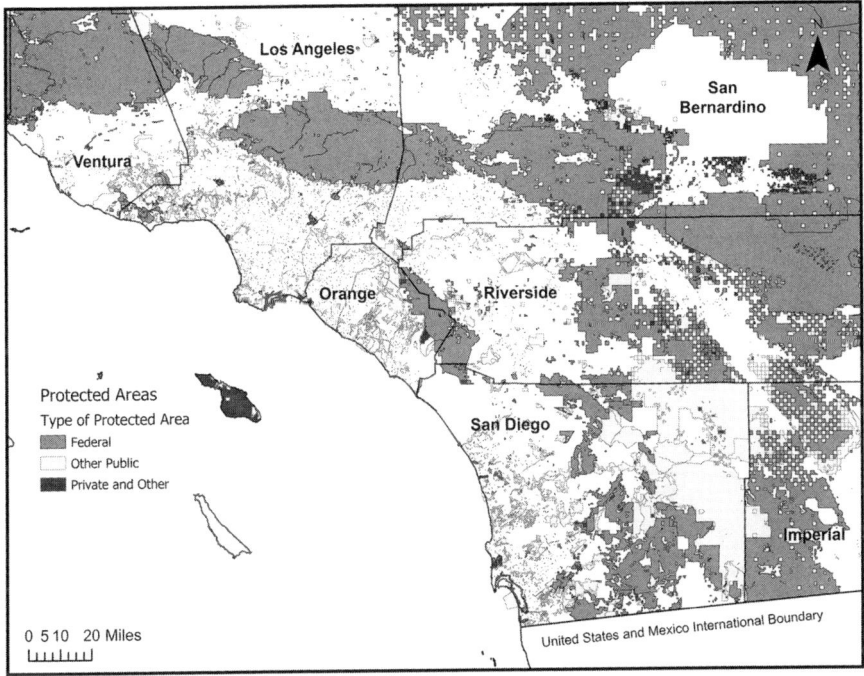

Figure 5.1. Map of public and private protected areas in Southern California. Most of the large protected areas are federally owned national forests in the mountains and national parks in the desert. County, city, and municipal protected areas tend to be smaller and isolated. Private protected areas, including land conservancy properties, are rare and usually quite small.
(California Protected Areas Database 2019; Basemap source: National Land Cover Database, 2011; Historical Land-Cover Change and Land-Use Conversions Global Dataset [1970, 1990]; UI-UC/ATMO, Department of Atmospheric Sciences, University of Illinois at Urbana-Champaign. Map prepared by Jessica Alger.)

Tennessee Valley Authority with federal funds. After the Supreme Court and the God Squad sided with the fish, Congress specifically exempted the Tellico Dam from ESA restrictions; Representative Al Gore voted for the dam, and Representative Newt Gingrich voted for the fish. Ecologists from the University of Tennessee hurriedly caught and transplanted all the snail darters they could find to another river system as the dam closed and the valley flooded, destroying all of the snail darter's known habitat. At present, the snail darter can be found in multiple streams in the area and is listed as threatened.[36]

Section 9—Take

Section 9 details all of the acts under the ESA that are prohibited without a permit, including killing or harming individuals of a listed species, harassing them or interfering with their breeding or feeding behaviors, importing or exporting them, keeping them in captivity, marking them for scientific study, or moving them around. Section 9 extends take prohibitions to include the destruction or degradation of occupied habitat and describes specific actions that can be interpreted as harassment. Unlike Section 7, which doesn't distinguish among taxonomic groups, Section 9 affords more protection to animals than to plants. The Services review activities to determine if they rise to the level of a Section 9 violation. If so, then the individual, company, or agency must seek a permit to conduct that activity.[37]

Court cases have heavily influenced the interpretation of Section 9 and the concept of take. Take is not limited to death or direct harm but can include degradation of occupied habitat to the point that an individual or population can no longer reproduce or perform other necessary behaviors there. Habitat destruction is covered under Section 9 if it directly causes harm to individuals of a listed species— destruction of unoccupied habitat is not take if it isn't designated as critical habitat, even if it may be beneficial for species recovery. In contrast to Section 7, which is focused on federal property and actions of federal agencies, Section 9 applies take prohibitions regardless of where it occurs and who causes it, whether they be individuals, organizations, or corporations. For example, state and local governments can be held liable if their ordinances or licensing generate take of

THE GNATCATCHER AND THE ESA

listed species, as illustrated in cases involving the permitted use of gill-nets and lobster pots that harm listed marine species that get caught in them, and the permitted use of motorized vehicles and outdoor lighting on public beaches used by listed species as breeding habitat.[38]

Private landowners who wish to use their land without restriction have a particularly adversarial relationship with Section 9, because take can include necessary activities such as clearing vegetation to build a new structure or disking weeds to reduce fire risk. In fire-prone Southern California, Section 9 has created conflict with private landowners who wish to clear flammable brush and grass away from structures. For example, landowners protecting their homes from fire in Riverside County were told that they could not disk the shrubs and grasses into the soil around their structures, because disking posed a risk to the endangered Stephens' kangaroo rat (*Dipodomys stephensi*), which lives in burrows. However, landowners were allowed to remove the vegetation by mowing it, which would not destroy the burrows. (More about this species in chapter 6.) The landowners' confusion over allowable brush-clearing methods generated animosity toward the kangaroo rat, as landowners perceived that the ESA placed the welfare of the kangaroo rat over that of people.[39]

Section 10

Section 10 allows the Services to permit exceptions to Section 9 for projects and activities that don't have a federal nexus. Before the 1982 amendment, only scientific collecting and conservation activities could receive Section 10(a) permits. The 1982 amendment added economic activities to the permitting process, opening the door to development-oriented Section 10(a) permits and HCPs. The motivation of this expansion was to reduce the ESA's economic impact on private landowners.

Section 10(a) permits are incidental take permits. Activities eligible for a 10(a) permit are those that do not intend to kill, harm, or harass a listed species—incidental does not refer to the level of impact to a species, which can be small to considerable, but rather to the intent of the activity. Incidental take permits are contingent on an approved HCP, which dictates actions for how the take will be mitigated. Every permit application is subject to a minimum thirty-day public comment period. As of 2016, the USFWS had issued 10(a) permits for almost 60 percent of potential gnatcatcher habitat.[40]

Section 10 also includes the No Surprises and Safe Harbor rules. The Services created these rules in the 1990s to soften political resistance to the ESA. Although not new—similar language was used in the first HCP at San Bruno Mountain in the 1980s—the rules explicitly reward private property owners who participate in approved HCPs with regulatory certainty in the future. The USFWS in particular was hoping to curtail the private property owner's 3-S approach to dealing with the ESA: Shoot, Shovel, and Shut up. As I describe in more detail in chapter 6, these assurances shift the risk of conservation uncertainty onto the federal government. Although not perfect, these assurances have improved landowner compliance with the ESA and have incentivized innovative conservation approaches by public-private partnerships.[41]

The Safe Harbor rule assures landowners that they will not require additional 10(a) permits if other listed species are attracted to the property after they engage in conservation activities such as habitat restoration. The Safe Harbor concept was designed by ecologists at the Environmental Defense Fund and adopted by the USFWS as a policy in 1995.

The No Surprises rule assures landowners that the activities and mitigation requirements covered in their 10(a) permits will be the final agreement; unforeseen circumstances cannot require landowners to engage in additional mitigation actions. Therefore, if a landowner follows the permit plan but the mitigation of harm to the listed species is unsuccessful and puts the species at risk, the USFWS then assumes responsibility and must acquire land elsewhere to protect or perform other activities to safeguard the species. The potential for failure increases when No Surprises assurances are issued without sufficient funding and data for land acquisition and monitoring. After the No Surprises rule began in 1994, the rate of proposed and approved HCPs increased considerably. The No Surprises rule and the listing of the coastal California gnatcatcher together provided enough incentive for landowners and developers to participate in the NCCP policy.[42]

As with just about every other ESA section, the coastal California gnatcatcher was involved in a legal challenge regarding the No Surprises rule and the conditions under which the USFWS could revoke a 10(a) permit. The plaintiffs, including environmental groups and Native American tribes, argued that the No Surprises rule violated the Administrative Procedures Act due to the lack of a public comment period. They successfully argued that No Surprises assurances represented a decision by the Services and, as such, must follow the public comment procedure. In response to the case, the Services revoked, rewrote, and opened the revised No Surprises rule to public comment and then adopted the revised version.[43]

Habitat Conservation Plans emerged from an urban conservation challenge in the 1980s: how to preserve two endangered butterfly species dependent on the last open space on the San Francisco Peninsula. Since their inclusion in the 1982 ESA amendment, HCPs have been criticized for their single-species focus, weak integration with recovery plan activities, lack of Services staff to review them, and inconsistent criteria used to measure their success. The use of HCPs in species conservation efforts was slow to catch on: only fourteen HCPs had been approved by 1992, and most of them were small, single-owner plans that did little to reduce threats across a species' range. The NCCP policy was a direct response to several catastrophic political clashes over listed species that had occurred in Southern California during the era of single-species HCPs.[44]

Habitat Conservation Plans go hand-in-hand with 10(a) permits. If you want a permit to conduct an activity that destroys habitat occupied by a listed species, you need an incidental take permit, which requires an approved HCP. Habitat Conservation Plans are usually written by professionals at local municipalities or consulting firms, depending on whether the landowner seeking the HCP is a public or private entity. The Services are obligated to approve an HCP and grant incidental take permits if the HCP meets the following criteria: the activities resulting in take are lawful and incidental; take will be minimized and mitigated to the maximum extent possible; the plan includes an adequate funding mechanism; the permitted take will not jeopardize the species' survival and recovery; and the plan will meet any other criteria set by the Service at the time of approval. All HCPs include an Implementation Agreement between the Service and the party(ies), and a steering or management committee to file paperwork, oversee reserve management and biological surveys, and manage funding for reserve support. Although not required, many HCPs involving multiple private landowners have some mechanism to collect mitigation fees, where a fixed amount is paid into an HCP fund for each unit area of habitat destroyed. These mitigation funds are used to purchase land elsewhere to conserve or restore habitat if required by the HCP and to perform such management activities as prescribed burns and monitoring. The outcome of an HCP is a habitat preserve for the listed species, and the permittee is responsible for monitoring and management activities for the duration of the permit.[45]

Organizing, designing, and managing an HCP requires considerable effort, resources, and expertise. Bringing more landowners to the table often brings economies of scale; more habitat can be preserved over a larger area for longer periods of time. Therefore, larger HCPs are generally better able to stabilize populations of listed species. But as the number of participants in an HCP grows, reaching consensus becomes difficult, and bureaucratic inertia can develop. Overly complicated HCPs can delay critical decisions or, worse, encourage hasty decisions that require more effort to fix later. Habitat Conservation Plans often have to be designed and approved before all of the necessary information is available; predictive habitat models can be used to fill these gaps but are only as good as the underlying data.[46]

As of 2016, there were almost a thousand approved HCPs in the United States covering roughly 40 million hectares, with an average permit duration of about twenty years. Over 70 percent of HCP permit holders were private individuals or corporations, and a quarter of all HCPs covered multiple listed species—these are called Multiple-Species HCPs, or MSHCPs. Habitat Conservation Plans can boost the probability of species recovery by 30–60 percent, and HCP reserves provide other benefits such as valuable open space for recreation, carbon sequestration, and flood protection. Remember that first HCP in San Francisco to protect the habitat of two endangered butterflies? It is now known as San Bruno Mountain State Park, where more than sixty thousand visitors a year take in stunning views of downtown San Francisco and the San Francisco Bay.[47]

ESA Success

ESA success lies in the eye of the beholder. Some expect that the nearly fifty-year-old law should have solved the endangered species problem by now. Conservative politicians and business-oriented special interests correctly argue that many listed species have languished for decades in the same status—including the coastal California gnatcatcher—while few have been recovered and delisted even after billions of dollars have been spent. Often, the Services do not have enough information on population sizes or trends of listed species to know whether they are recovering. In the USFWS biannual "Reports to Congress" regarding listed species recovery, the status of listed species is often based on qualitative information and the best guess of a species expert.[48]

Scientists and environmentalists have a far rosier view of the ESA. The act has slowed the decline of at-risk species and provided incentives to protect and restore habitat. The number of observed extinctions in the United States is lower than what we should expect, statistically speaking, than if the ESA had never been enacted. More species go extinct in the candidate-for-listing stage than those that are listed. For both marine and terrestrial species, the longer a species has been listed under the ESA, the better it has fared. Four or five decades may not be enough time for some species to recover, particularly for large-bodied species that reproduce slowly and for species that require a large number of individuals to breed successfully. This is a good place to note that the famed recovery of the bald eagle took almost thirty-five years. The majority of the currently listed species have only been protected by the ESA for twenty years or less and were closer to extinction when they were listed than species like the bald eagle. These numbers recommend patience.[49]

Bird conservation has been particularly successful under the ESA, which has stabilized or aided the recovery of 85 percent of all listed bird species. Populations of ESA-protected birds generally stabilize by five to ten years after listing, then increase, boosting the likelihood of recovery and delisting. A 2016 study by the Center for Biological Diversity estimated that populations of ESA-listed bird species increased by almost 650 percent between 1968 and 2014, with dramatically better long-term outcomes than nonlisted species.[50]

Are specific sections of the ESA responsible for these results? Species are more likely to experience population increases if they have critical habitat designated and a recovery plan in place, although support for the effectiveness of critical habitat designations is murky. Successful outcomes for listed species—from halting population declines to recovery and delisting—are positively related to funding. Pernicious funding shortages diminish the likelihood of successful recovery because the shortages prolong the time it takes to move a candidate species into a listed status, delaying needed conservation actions. Some of these funding shortages are due to political influence on conservation priorities. Just 5 percent of listed species receive 80 percent of the Services' funding designated for ESA enforcement, with fifteen species of salmon and sturgeon receiving the majority of that 80 percent. Unfortunately, the USFWS's budget has been declining since 2010, creating even more dire funding conditions as the number of ESA-listed species grows.[51]

All would agree that the ESA has not recovered as many species as is desirable. Even with an unlimited amount of money, the ESA would still struggle to counter the strengthening threats that place species in peril: habitat loss, overharvesting, invasive species, climate change, and pollutants. Given that these threats have not abated, delisting may be impossible for some species and therefore should not be used as the sole criteria for ESA success. For species characterized by a few hundred individuals restricted to small ranges under constant threat, delisting may never be possible.[52]

It should be no surprise that ideas to improve the ESA emanate from many corners. California's NCCP approach was created to address some of the shortcomings of the ESA: its antagonistic relationship with private property; its reactive, single-species, project-by-project focus; its bias toward well-known and appealing species; and its inadequate funding and resources. The NCCP policy also seeks to incentivize private landowners who actively conserve biodiversity as part of a regional conservation network. Incentives include free technical assistance to landowners for habitat restoration, tax abatement for enrolling high-value habitat into an HCP, and direct payments on a per-hectare basis through a tradable mitigation permit system. Tradable permit systems are being used in the NCCP policy to compensate private landowners for conservation on their property and to locate development away from valuable conservation areas. The NCCP policy is one of several affiliated with the ESA that aim to make conservation more proactive and cost-effective. Other policies include Candidate Conservation Agreements, Safe Harbor Agreements, and a Prelisting Conservation Policy.[53]

6

NCCP to the Rescue

Biodiversity conservation in fire-prone and urban-sprawled Southern California has long been challenging. For decades, the specter of Endangered Species Act listings has shadowed the building industry as it provided housing, offices, and shopping to a growing population. Government revenue ceilings due to Proposition 13 and a white-hot housing market restrict the ability of local governments to purchase land for species conservation and open space. Land-use planning, private property rights, and economic incentives driving sprawl are at the heart of most conservation issues in California and inspired the Natural Community Conservation Planning approach.[1]

When it was enacted in 1991, Interior Secretary Bruce Babbitt called California's NCCP policy "unprecedented," "trailblazing," and "pioneering," "an example of what must be done across the country if we are to avoid the environmental and economic train wrecks we've seen in the last decade." At the time, Secretary Babbitt was facing active attempts by congressional Republicans to eviscerate the ESA, a response to the growing outcry from private landowners over ESA restrictions on their property (see chapter 5). Secretary Babbitt hoped that this new, voluntary approach to species conservation would quell the uproar.[2]

The NCCP policy's goal is to allow for continued development while protecting biodiversity by moving away from the ESA's species-by-species, project-by-project approach. Shifting the focus of con-

servation to habitats and regions, the NCCP policy encourages public and private landowners to cooperatively set aside large networks
of protected habitat to support many species at once, stabilizing
species numbers and eliminating the need for further ESA listings.
The NCCP approach relies on local municipalities to coordinate with
each other on land-use planning and reserve-network design and to
seek buy-in from important stakeholders. In the words of the California Department of Fish and Wildlife: "Through local planning, the
NCCP program aims to protect wildlife and habitat *before* the landscape becomes so fragmented or degraded by development that the
listing of individual species is required."[3]

At the start of its implementation, the state of California hoped
that the NCCP policy would prevent future conservation fiascos,
such as the one that erupted in the late 1980s with the Stephens'
kangaroo rat in Riverside and San Bernardino Counties. Governor
Pete Wilson appointed two World Wildlife Fund employees, Douglas
Wheeler and Michael Mantell, as secretary and undersecretary of the
California Resources Agency (respectively) to solve the state's growing conflict between development and the ESA. Wheeler and Mantell were the chief architects of the NCCP policy, with significant
input from stakeholders such as the Irvine Company, a large Orange
County developer. For the federal government, the NCCP policy was
a "proving ground" for the shift from the single-species Habitat Conservation Plans to Multiple Species HCPs under the ESA.[4]

As the NCCP process gained steam, many were cautious about its
outcome. In 1994, Jon Atwood and Reed Noss stated: "In 10 years, we
may find that what seemed a perfectly adequate plan for conserving
coastal sage scrub was, in fact, not strong enough to maintain viable
populations of gnatcatchers. Alternatively, we may find that we protected more than enough land for gnatcatchers but not enough (or
not in the right places) for several other species, which are now imper-

iled enough to qualify for listing under the ESA. Or perhaps we will find that we guessed everything just right." Six years later, Atwood wrote, "Regulators and policy makers 10 years from now may be faced with the embarrassing possibility that, despite thousands of hours and millions of dollars spent toward planning efforts, still no one is sure if the NCCP's flagship species [the coastal California gnatcatcher] has been successfully conserved—or if it is declining, what to do about the problem."[5]

Now thirty years after the NCCP approach began, the question still remains: Is the NCCP policy working?

The Nuts and Bolts of the NCCP Policy

The NCCP Act was authorized by California law AB 2172 on October 10, 1991, and amended in 2003 to add a requirement for "adaptive management" in NCCP plans. The NCCP was part of Governor Pete Wilson's Resourceful California initiative, a broad set of programs focused on protecting and preserving the state's environmentally and culturally important sites. The NCCP Act gives the California Department of Fish and Wildlife the authority to negotiate implementation agreements with participants and to issue guidelines for reserve design and other plan elements. The coastal California gnatcatcher's ESA listing in 1993 created an incentive for landowners to join an NCCP plan, rather than go through the ESA HCP process, if they wanted to destroy or modify gnatcatcher habitat. The U.S. Fish and Wildlife Service's decision to list the gnatcatcher as threatened—rather than endangered as the agency originally planned—allowed the USFWS to issue a 4(d) special rule for the gnatcatcher, which permitted incidental take of gnatcatchers and their habitat while NCCP plans were being developed. Note that this was different than the issuance of incidental take permits associated with an HCP; those permits are not issued until after the HCP has been approved. The 4(d) special rule also delegated authority for ESA enforcement from the federal government to the state of California. These two decisions thus paired the fate of the NCCP policy to that of the gnatcatcher, even though the intent of the policy was to move beyond the single-species approach. The gnatcatcher listing under the ESA "fundamentally altered the NCCP program, grafting the regulatory power of FESA [the Federal ESA] onto the state's heretofore voluntary habitat planning process."[6]

Currently, the NCCP process starts with a biological resource assessment, then moves to planning and reserve design and approval of incidental take permits, and then to implementation agreements. Local governments develop NCCP plans, which must be approved by both the USFWS and the California Department of Fish and Wildlife, and then private landowners enroll their property as third parties in the local government's plan. Once an NCCP plan is approved, participating landowners can receive incidental take permits for many listed species at once. Outside of an NCCP plan, landowners who want to develop land with listed species must work directly with the USFWS on an HCP or MSHCP.[7]

NCCP plans often build on existing systems of public nature preserves and designated open space. The plans expand and connect these areas using land acquisition and land swaps, conservation easements on private land, and habitat mitigation banks based on tradable development rights—also called conservation banks. In fact, the NCCP policy was viewed by federal and state agencies as an experiment in the use of mitigation banks for endangered species. Mitigation permits and easements provide a mechanism to compensate private landowners who volunteer to protect habitat on their property, allowing habitat to be altered or destroyed by development in other areas. These habitat banks have had mixed success under the NCCP policy. The first habitat bank created was the 191-hectare Carlsbad Highlands preserve for coastal sage scrub and the California gnat-catcher. Today, the preserve suffers significant damage from off-trail use, generating conflicts between state game wardens patrolling the area and mountain bikers using the area illegally. In another NCCP area, the San Bernardino Valley Audubon Society successfully sued the Metropolitan Water District to prevent the use of a banking tool for the Lake Mathews MSHCP/NCCP plan. While mitigation banks are commonly used around the country to preserve wetlands, similar banks for ESA-listed species operate at smaller scales.[8]

One of the most important differences between a federal MSHCP/HCP and a state NCCP plan is the level of protection expected from them. A federal HCP requires only that harm to a species be minimized and mitigated. While the NCCP Act similarly requires that NCCP plans focus on species "conservation and management," the California Department of Fish and Wildlife has interpreted this clause to include species recovery, and thus a higher standard of protection. Nevertheless, the NCCP policy is not an alternative to ESA enforcement. The ESA is a mandatory regulation that affects all landowners and can be enforced by the federal government. The NCCP policy is voluntary, affecting only those landowners who choose to enroll in NCCP programs. NCCP participants can back out of the program at any time with little penalty, reverting back to the 10(a) permit system under the ESA.[9]

The initial overselling of the NCCP policy as the solution to all the ESA's problems damaged its credibility among scientists and conservationists. Although voluntary environmental programs for private landowners have become extremely common in the United States, measuring the effectiveness of these programs is rare and program participation is often stubbornly low. As discussed in chapter 5, the USFWS decided not to delineate critical habitat at the time of the coastal California gnatcatcher's listing because the Service assumed that the habitat protected under the NCCP policy would be sufficient to preclude any further decline in the species. In 1997, the Natural Resources Defense Council forced the USFWS by lawsuit to designate critical habitat for the gnatcatcher. The NRDC successfully argued that, as a voluntary program, the NCCP policy could not take the place of legally enforceable critical habitat designations.[10]

The NCCP policy's promises of regulatory relief and bureaucratic streamlining were a huge selling point for private landowners. A few years before the gnatcatcher's listing, the Stephens' kangaroo rat was listed as endangered under the ESA. That listing hobbled the development industry in Riverside and San Bernardino Counties. Derisively called the "K-rat" by locals, the Stephens' kangaroo rat is a large-headed, nocturnal rodent that has a distinctive hopping movement due to its elongated hind legs, similar to kangaroos. It is roughly 30 centimeters long from nose to tail tip and lives in burrows in grasslands and grassland–coastal sage scrub mixtures. Agriculture and development are major threats to the rodents' existence, since both disturb or destroy the top layer of soil where these animals construct their burrows.[11]

By the time of its listing in 1988, the kangaroo rat's habitat had been reduced by two-thirds and severely fragmented, with only 6 percent of its remaining habitat under any protection. Most of this remaining habitat was on private property, and the amount of habitat in public ownership was insufficient to ensure the species' survival. Developers and landowners grabbed headlines with their protests over the delays and expense due to the species' listing, inspiring "Steve, you dirty rat!" bumper stickers. At one point, the entire Riverside County Board of Supervisors faced possible jail time for approving new development projects in kangaroo rat habitat without an approved HCP. As with the coastal California gnatcatcher, the kangaroo rat's listing has been repeatedly challenged by the Southern California building industry.[12]

One lesson that the state of California learned from the Stephens' kangaroo rat experience was to avoid complete development moratoriums. Moratoriums create havoc for the building industry and uncertainty for landowners, generating ill will toward listed species in particular and biodiversity conservation in general. The NCCP process is designed to save developers and landowners time and uncertainty. Dealing with local governments under an NCCP plan reduces the bureaucracy that landowners must navigate, and the Safe Harbor and No Surprises clauses granted to approved NCCP plans reduce the risk and uncertainty landowners face. The NCCP policy therefore offers a substantial incentive to landowners to participate in the program, particularly if their land is home to multiple species listed under the ESA.[13]

Early Stumbles

The NCCP policy had a wobbly start. Before the gnatcatcher's listing in 1993, the coastal sage scrub pilot NCCP program had enrolled just thirty-seven private landowners and thirty-three municipal and county governments, representing about half of the remaining coastal sage scrub in the area. Given that a nontrivial portion of the enrolled land was to be developed, that level of program participation was not likely to preserve enough coastal sage scrub to preclude future ESA listings. Worse, in 1992 the state legislature threatened a 75 percent cut in funding to NCCP programs, making it impossible for state and local agencies to collect the background data needed to develop science-based conservation guidelines. Although the USFWS's listing of the coastal California gnatcatcher in 1993 was meant to bolster the NCCP policy, not everyone saw it that way.[14]

In May 1993, President Bill Clinton was confronted at a public forum in San Diego over the negative impact that the coastal California gnatcatcher's recent listing was having on the local construction industry. At the time, Southern California was facing a severe recession, with high unemployment in the construction and housing sectors. The president responded, "Just north of here, I thought the Secretary of the Interior had made an agreement that allowed construction to go forward.... I thought the deal that he hammered out on the gnatcatcher up north ... would have general application and would stop this kind of problem."[15]

President Clinton was likely referencing the USFWS's controversial Section 7 "no jeopardy" decision under Interior Secretary Babbitt, which allowed the construction of the San Joaquin Hills tollway, a portion of State Road 73. As a transportation project receiving federal funds, the project required a Section 7 consultation with the USFWS, which determined that the project would not jeopardize the coastal California gnatcatcher's long-term survival. The tollway sliced through 63 hectares of occupied gnatcatcher habitat and was the largest remaining tract of open space in Orange County. The USFWS's no jeopardy decision was not made in isolation. In 1989, Robert Thornton's Irvine-based law firm—the same firm that later represented the building industry's lawsuits against the gnatcatcher's listing—was placed on retainer for the Orange County Transportation Corridor Agencies. On behalf of the transportation agencies, Thornton began negotiating with the USFWS to issue a no jeopardy decision for all the listed species in the path of the San Joaquin Hills tollway. The Service's decision encouraged Orange County developers to stick with the NCCP program and allowed Governor Wilson to continue to support the NCCP policy in a tough political climate.[16]

The NCCP policy's brush with death in 1996 was less public but no less serious. When the USFWS issued the 4(d) special rule for the coastal California gnatcatcher in 1993, the agency allowed the state to issue incidental take permits for gnatcatchers while NCCP plans were in the planning stage. The USFWS set these permitted losses at no more than 5 percent of existing habitat or 116 gnatcatcher pairs, whichever came first. Once those limits were reached, no new incidental take permits could be issued through the NCCP process until plans were approved. The 5 percent habitat loss limit represented 8,095 hectares of the 162,000 remaining hectares of habitat, and 116 pairs was roughly 5 percent of the estimated 2,562 gnatcatcher pairs in the United States thought to exist at the time. These losses were supposed to come from areas with low conservation value and were to be mitigated via the NCCP process. But this didn't happen. By 1996, although the cumulative take of coastal sage scrub was under the limit, the cumulative take of gnatcatchers exceeded 116 pairs. In response, the Spirit of the Sage Council, a regional Native American tribal organization, filed a sixty-day notice of its plan to sue the USFWS, and ultimately did.[17]

Jolted into action, the USFWS reexamined its Section 7 consultation and the 4(d) special rule for the gnatcatcher. The Service collated surveys that had occurred since the listing and revised its population estimate upward to 3,430 gnatcatcher pairs. The revision increased the allowable take limit up to 170 pairs, enough to cover the additional take requests anticipated in Orange, San Diego, and Los Angeles Counties. The USFWS attributed the increased population estimate to "surveys occurring within previously unsurveyed areas and increased numbers of birds due to variations in productivity of sage scrub in response to climatic conditions ... [but] should not be construed to represent an increase in the overall gnatcatcher population." Once the NCCP program gained momentum and regional plans were approved, the USFWS lifted the interim take limits.[18]

The Coastal Sage Scrub Pilot Program

Although the NCCP Act was written to preserve all habitat types, the California Department of Fish and Wildlife first targeted the severely reduced coastal sage scrub community as a pilot program. The hope was that the NCCP process would protect enough coastal sage scrub to avoid further species declines and additional ESA listings. At the time, most unprotected coastal sage scrub was privately owned, representing some of the priciest real estate in the world.[19]

The 15,500-square-kilometer NCCP planning area in Southern California was divided into eleven subregions, dictated primarily by political boundaries (such as city and county lines). Using political boundaries rather than biological ones simplified the bureaucratic process, but doing so exacerbated the lack of coordination among plans and ecological connectivity among NCCP preserves. The 5 percent interim take limit on coastal sage scrub applied to each subregion; once the subregion's NCCP plan was approved, the take limit was lifted.[20]

Although a diversity of stakeholders and interest groups drive the NCCP process, scientific evidence is supposed to govern the design and implementation of NCCP habitat preserves. Ideally, NCCP habitat networks should meet conservation science ideals: conserve all species throughout the entire plan area; aim for large reserves that are well connected and proximate, rather than many small, isolated reserve fragments; develop a portfolio of reserves that will provide the widest spectrum of physical and environmental conditions; and minimize human impact in the reserves. Unfortunately, science has not always maintained its prominent spot in the process. Mechanisms to ensure scientific influence that were prominent in the early years of the pilot program quickly fell by the wayside.[21]

For example, the NCCP Scientific Review Panel was an important—albeit brief—part of the pilot NCCP process. Chaired by Dennis Murphy, who was the director of Stanford University's Center for Conservation Biology, the panel included other well-known and respected conservation biologists—Reed Noss, Michael Gilpin, Peter Brussard, and John O'Leary. The panel drafted a set of guidelines for preserve network construction and identified target species to use as indicators of conservation success: coastal California gnatcatcher, coastal cactus wren (*Campylorhynchus brunneicapillus*), and a lizard called the orange-throated whiptail (*Aspidoscelis hyperythra beldingi*). The panel chose the California gnatcatcher because its range generally co-occurs with the range of coastal sage scrub at greatest risk of loss, it is nonmigratory and sings year-round (advantageous for surveying), and it had documented population declines. At the beginning of the pilot program, there were only 2,562 pairs of California gnatcatchers and 138,807 to 179,700 hectares of their habitat left in the United States—about 14 to 18 percent of the historic extent. Less was known about the cactus wren and orange-throated whiptail, and the panel emphasized the importance of species surveys, life history information, and genetic studies to the panel's work. Unfortunately, none of this information was forthcoming—most of it was still not available in 2020—and ultimately the panel disbanded in 1993. The NCCP process that was to be driven by science instead drifted into a "quasi-political, quasi-scientific process . . . infused by practicality." Policy makers and politicians feared that they would lose their much-needed buy-in from developers and landowners if they waited too long for the data that scientists needed. Therefore, they chose to move the planning process forward, sacrificing scientific certainty for political support.[22]

Early on, the development industry supported the NCCP policy since it provided them with regulatory certainty. Participating developers could set aside some habitat on the project site or elsewhere, and the rest of the property could be developed without additional conservation measures imposed in the future, thanks to the No Surprises assurance. But enthusiasm for the NCCP policy was uneven. The first NCCP plan in Orange County was quite popular—early enrollees represented 90 percent of the county's coastal sage scrub, due to the participation of the Irvine Company and a few other large landowners. In other counties, early enrollment was anemic. Meanwhile, environmental groups were split in their support of the policy. The Nature Conservancy supported it, but the Natural Resources Defense Council, refusing to drop its petition to list the gnatcatcher, was concerned about the voluntary nature of the NCCP policy and the automatic inclusion of the No Surprises assurance. If the NCCP policy failed, the USFWS would be required to enact the necessary mitigation measures, likely hampered by inadequate resources. Soon after the NCCP policy was operational, the Spirit of the Sage Council sued the USFWS over the No Surprises assurance, arguing that the NCCP policy's promises to developers contradicted protection requirements in the ESA.[23]

In the first two years of the pilot program, it became clear that most small landowners had insufficient incentive to enroll their property in an NCCP plan without the threat of listed species and ESA enforcement. While large landowners had greater incentive to participate, they did not own enough land on their own to set up protected habitat networks without the participation of at least some of the smaller landowners. The many, small habitat patches owned by small landowners were necessary to connect the preserve networks set up by large landowners. But small landowners simply waited for the large landowners and others to enroll enough habitat to satisfy the NCCP reserve requirements. This was problematic for the NCCP policy's forward momentum. Worse, the petitions to the USFWS from Atwood and the NRDC to list the gnatcatcher set off a wave of habitat destruction; from 1991 to 1992, over 800 hectares of coastal sage scrub were destroyed. Landowner enrollment remained low until the California gnatcatcher's listing in 1993.[24]

In addition to providing an incentive to landowners to participate in NCCP plans and avoid the ESA permit process, the gnatcatcher's listing transformed the NCCP policy from a wholly state effort to a collaboration between the state and federal governments. While seeming to contradict the intent of the NCCP policy to eliminate new ESA listings, the USFWS made it clear that the gnatcatcher's listing was not a no-confidence vote for the NCCP policy. At the time, the director of the USFWS recognized that "without the listing of the [gnatcatcher], the framework of incentives and compromise on which the NCCP [policy] is based is imperiled." When a federal district court judge overturned the Service's decision to list the coastal California gnatcatcher on procedural grounds in May 1994, he allowed the gnatcatcher to stay listed to help preserve the progress made by the NCCP policy while the USFWS corrected a procedural error in the listing process (see chapter 5). Once the NCCP process gained momentum, the USFWS signaled that it would not accept single-species HCPs that did not conform to NCCP standards, further encouraging landowners to work through the NCCP process. These were clear signals that the federal government actively supported the NCCP policy.[25]

A deeper dive into how well each county proceeded with the NCCP policy can be instructive: Which counties did better, and why? At the start of the NCCP process, the pilot region was divided into eleven subregions that were meant to develop their own NCCP plans; no subregions included habitat in more than one county. Orange and San Diego Counties emerged as early trend setters in establishing NCCP plans. By 1995, both counties had enrolled almost all eligible coastal sage scrub in NCCP or NCCP/MSHCP plans, totaling 40,208 hectares. Three subregional NCCP plans covered 95 percent of gnatcatcher habitat in San Diego County. By 2011, four of the original eleven planned subregional NCCP plans had been approved; the large Orange County Coastal/Central NCCP plan and the San Diego NCCP/MSCP plan, as well as two smaller, narrowly focused utilities' NCCP plans.[26]

In 2010, the USFWS reported that NCCP plans were collectively on track to conserve almost 60 percent of the gnatcatcher's U.S. habitat once the NCCP plans were fully implemented. The USFWS later revised this estimate downward to about 55 percent, based on new vegetation data and gnatcatcher habitat models that identified more unprotected habitat. By 2016, the numbers were revised further downward to about 51 percent of suitable gnatcatcher habitat protected to some degree; 16 percent was permanently protected with minimal uses allowed, and 35 percent was permanently protected with allowable uses, such as driving off-road vehicles and mining. These percentages do not include all strictly protected and multiple-use protected areas outside of the NCCP process, such as pre-NCCP designated state parks and other natural areas. Approved NCCP plans received permits for up to seventy-five years and, once fully implemented, would preserve nearly 75,000 hectares of gnatcatcher habitat. Large federally owned areas, particularly the coastal military bases of Camp Pendleton and Miramar, the Cleveland National Forest, and the San Diego National Wildlife Refuge, were expected to provide some connectivity among the NCCP subregional plans.[27]

Over the past thirty years, on average one hectare of coastal sage scrub has been approved for loss for every hectare placed into an NCCP habitat reserve. Although overall habitat amount has declined, the USFWS maintains its optimism that current losses will be mitigated by future gains. The picture south of the U.S.-Mexico border remains fairly bleak. Although two private nature reserves were established on the Baja California Peninsula, these reserves encompass little coastal sage scrub. Coastal sage scrub remains under threat from residential and commercial developments, particularly related to tourism, grazing, and agriculture. Mexico does not have a law as strong as the ESA that would protect California gnatcatchers outside of protected areas, and thus, as of 2020, the coastal California gnatcatcher remained unprotected throughout most of its range outside of the United States.[28]

Current Status of the NCCP Program

As of December 2019, there were sixteen approved NCCP plans in California incorporating over 1.5 million hectares, which when completed will create a reserve system of 637,000 hectares for hundreds of species in all kinds of habitats. Another nine NCCP plans are in the planning process, incorporating 1.6 million hectares. When all plans are completed, almost 9 percent of the land area in California will be covered under the NCCP policy. The majority of these NCCP plans are in Southern California, where eighteen NCCP plans are in the planning or approval stage—Ventura County is the only one of the six counties within the California gnatcatcher's range that does not have an NCCP plan in any stage of the process (table 6.1).[29]

Although the NCCP process remains dependent on development impact fees for funding land acquisition and management, public and nonprofit participants can apply for funding through the state's NCCP Local Assistance Grant program to support management plans, land acquisition, monitoring, and research. Additionally, several plans have developed novel funding sources or participation incentives. For example, the Western Riverside MSHCP/NCCP plan requires local agencies seeking funding from a transportation bond to participate in the NCCP plan, incentivizing land enrollment. The city and county of San Diego use general fund resources, sales

Table 6.1 Statistics for NCCP plans in Southern California, as per 2019[a]

County[b]	Plan name[c]	Year approved	Permit duration (years)	Reserve system hectares (ha) total as of 2019 (hectares of CSS)	Species covered by plan[f]
Los Angeles	City of Rancho Palos Verdes NCCP/HCP	In planning	40	568 ha (298 ha)	10
Orange	Orange County Transportation Authority NCCP/HCP	2017	40	524 ha	12
	County of Orange Central/Coastal Sub-region NCCP/HCP	1996	75	15,126 ha (7,498 ha)	44
Riverside	Western Riverside County Multiple Species NCCP/HCP	2004	75	202,343 ha	146
	Coachella Valley Multiple Species NCCP/HCP[e]	2008	75	301,855 ha	27
San Diego	San Diego County Water Authority NCCP/HCP	2011	55	777 ha	63
	San Diego Gas and Electric Subregional NCCP/HCP	1995	55	97 ha	110

San Diego (cont.)	San Diego County Multiple Habitat Conservation Program (MHCP)	2004 1 subarea approved in 2004 (Carlsbad), 1 subarea in planning (Oceanside)	50	7,689 ha; 2,590 ha in Carlsbad subarea	85 (63 in City of Carlsbad)
	San Diego North County MSCP	In planning	50	—	62
	San Diego East County MSCP[d]	In planning	—	647,497 ha	157
	San Diego (South) County Multiple Species Conservation Program	5 subareas approved (1996–2005), 1 subarea in planning (Santee)	50	69,574 ha	85
San Bernardino	Town of Apple Valley MSHCP/NCCP[e]	In planning	—	17,806 ha (0 CSS ha)	50

[a] Data collated from individual plan documents in December 2019, available at https://www.wildlife.ca.gov/Conservation/Planning/NCCP/Plans.

[b] Ventura County had no NCCP plans in planning or approved as of 2019. The county was originally excluded from the NCCP pilot planning region for coastal sage scrub.

[c] NCCP = Natural Community Conservation Plan; MSCP = Multiple Species Conservation Plan; MHCP = Multiple Habitat Conservation Plan; MSHCP = Multiple Species Habitat Conservation Plan; HCP = Habitat Conservation Plan.

[d] The San Diego East County MSCP progress has slowed due to budget and staffing constraints and is not included in the December 2019 summary of NCCP plans.

[e] California gnatcatcher not included; area is too far east.

[f] A 2006 study of MSHCPs in California determined that many plans included species as "covered" in the plan that were not ever confirmed as present in the planning area. Matthew E. Rahn, Holly Doremus, and James Diffendorfer, "Species Coverage in Multispecies Habitat Conservation Plans: Where's the Science?" BioScience 56 (2006): 613–619.

tax revenues, and transportation project impact fees to help support their NCCP plans. Proposition 12, which passed in 2000, released $100 million to the state Wildlife Conservation Board and $50 million to the Department of Parks and Recreation that could be used for land acquisition for NCCP plans. These funds, along with other state and local sources, have provided between $1 million and $14 million per year in funding for land acquisition since 2000. Throughout the NCCP policy's history, land trusts and environmental organizations have been instrumental in acquiring large areas for conservation. Southern California benefits from the land preservation activities of dozens of nonprofit organizations, including many operating specifically in Los Angeles, San Diego, and Orange Counties.[30]

Despite the positive momentum for land acquisition, the connectivity of NCCP reserves across jurisdictional boundaries remains suboptimal: complexity of the NCCP process has generated dozens of poorly coordinated plans. Given this complexity, some have argued that streamlining the NCCP process would help improve enrollment within each NCCP plan and coordination across NCCP subregions. In particular, combining the NCCP permit process with other environmental permitting issues, such as wetlands protection under the Clean Water Act, might make sense. If a development project is slated to damage or destroy many types of habitats affecting many species, dealing with all the permits at one time, with one agency, is desirable.

A word of caution on that idea. Although bureaucracy and red tape can prolong permitting processes and add to the cost of planning and development projects, redundancy in governance reduces the risk of negative, irreversible outcomes. Just as redundant systems in an airplane reduce the likelihood of a crash, so redundant policies increase safety, reliability, and adaptability. These allow us to detect and correct errors before they accumulate through a system, causing an irreversible, catastrophic failure—such as a species extinction or a contaminated water supply. We should make sure these permitting requirements are not risk-reducing redundancy before we eliminate them.[31]

Details on the more prominent NCCP plans illustrate how these plans are designed and implemented, why some have been successful, and why others have stalled or failed. The most illuminating NCCP plans include the early adopters in Orange County, the slow and controversial Western Riverside County NCCP plan, the crazy patchwork of the NCCP plans in San Diego County, and the tiny NCCP plan on Rancho Palos Verdes in Los Angeles County. Each of these plans must contend with different legacies of land use and land ownership, funding restrictions, and landowners choosing to work with or against the process. In some cases, the process proceeds faster and better than anyone could predict—in Orange County, a large and sympathetic developer virtually ensured that the Central/Coastal NCCP plan would be a success. In other cases, the outcomes are less rosy—state and county agencies work tirelessly to acquire specific parcels of high conservation value, only to have the deal fall through in the end.

Orange County NCCP Plan

Among the Southern California NCCP subregions, Orange County achieved the earliest success for several reasons. First, Orange County had an advantage due to its history of land ownership that culminated in a few large owners rather than many small ones. Specifically, the legacy of large Spanish ranches resulted in a substantial proportion of Orange County ending up in the hands of just four families: Rancho Los Alamitos, owned by the Bixby family; Rancho Santa Margarita y Los Flores, owned by the O'Neill family; and two ranches—Rancho San Joaquin, owned by the Sepulveda family, and Rancho Santa Ana, owned by the Yorba family—that combined to form the Irvine Company developer. The Irvine Company, which owns or manages about a fifth of the county, had been working with the county government on open-space planning for two decades before the NCCP policy. When the Orange County Central/Coastal NCCP plan was approved in 1997, the NCCP plan had already set aside 88 percent of the coastal sage scrub (about 13,311 hectares) needed for the plan thanks to the Irvine Company, leaving just 1,815 hectares left to procure at an estimated cost of $8–$9 million.[32]

The second reason for the Orange County success relates to the first: the NCCP stakeholders had already realized that pairing open space with conservation made political and economic sense. Residents want open space for recreation and aesthetics, and houses near open space sell for more money. Furthermore, the public outcry over the loss of a large swath of occupied gnatcatcher habitat to the San Joaquin Hills tollway brought the Orange County Transportation Authority on board as a stakeholder, if only to avoid more bruising battles in courtrooms and the media.[33]

Orange County is home to two approved NCCP plans: the Orange County Central/Coastal NCCP/HCP plan and the Orange County Transportation Authority NCCP/HCP plan. Between 1992 and 1994, nineteen organizations signed NCCP enrollment agreements with one of the two plans: ten cities or other municipalities, the Transportation Corridor Agencies, Marine Corps Air Station El Toro, and seven companies, including several large landholders (Chevron USA, Irvine Company, and Santa Margarita Company). The early formation of a coalition of large landowners like the Irvine Company, environmental groups, and municipal governments facilitated the NCCP policy's early start and success in Orange County.[34]

The Orange County Central/Coastal NCCP/HCP plan gained USFWS approval in 1996 (table 6.1). The planning area covers 84,580 hectares, including about 42,000 hectares of coastal sage scrub in two large, unconnected blocks: one inland (the Central) and the other along the coast, which was split in two by the San Joaquin Hills tollway. The plan uses the three species recommended by the Scientific Review Panel for conservation and monitoring: the California gnatcatcher, the cactus wren, and the orange-throated whiptail. The total 15,126-hectare reserve system was assembled mainly from contributions by two large landowners: the Irvine Company, which contributed a 6,880-hectare preserve and later set aside another 4,775 hectares; and the Santa Margarita Company, which contributed 3,237 hectares. The reserves protect roughly 75 percent of the remaining coastal sage scrub in the county, 82 percent of known pairs of California gnatcatchers, and 77 percent of known cactus wren pairs.[35]

The county requires large development projects to set aside open space as well as pay development impact fees to support management and monitoring of the reserve system. Even development projects not enrolled in the NCCP plan are required to pay these fees. A unique feature of the Central/Coastal NCCP/HCP plan was the creation of the Natural Communities Coalition, a nonprofit organization established when the plan was approved in 1996 to coordinate the plan's 15,000-plus hectare reserve system. Funded by accumulated impact fees and sizable donations from the Irvine Company, the Nature Conservancy and the Irvine Ranch Conservancy conduct the long-term preserve monitoring and management activities. As the reserve network nears completion, NCCP activities have turned toward restoration efforts rather than habitat set-asides. The NCCP plan has also increased its focus on habitat for the coastal cactus wren–coastal sage scrub dominated by cactus.[36]

The Orange County Transportation Agency NCCP plan includes 26,550 hectares of coastal sage scrub hosting 3,419 individual gnatcatchers. As of 2020, 78 percent of these hectares were protected, supporting 66 percent of the gnatcatchers living in the planning area. Note here that "protected" does not mean protected by this specific NCCP plan. Indeed, this plan directly manages only 171 hectares of coastal sage scrub supporting about a dozen pairs of gnatcatchers, with plans to restore another 57 hectares to support another dozen pairs. How is this possible? The plan follows the long, linear road network managed by the agency, which bisects many areas that are already protected as state or county parks or owned by land conservancies. Thus, the Transportation Agency NCCP plan focuses on protecting or restoring areas that are critical habitat corridors and not already protected. Two examples of recent restoration projects are the Aliso Canyon Preserve in the San Joaquin Hills and the North Coal Canyon wildlife undercrossing for State Road 91 between the Santa Ana Mountains and Chino Hills.[37]

The Orange County Transportation Authority's NCCP plan is one of few that benefits from a funding source not tied to additional road building or development. Approved in 1990, County Measure M generated funding for transportation projects through a $0.05 local sales tax; the measure was reapproved as M2 in 2006. Five percent of this fund is designated for wildlife and habitat protection and mitigation. Generally, the Transportation Authority uses these funds for restoration projects or land acquisition. As of 2017, it had purchased seven high-conservation-value properties to set aside in preserves, for a total of 524 hectares. These properties were identified and recommended for acquisition by an environmental group, the Friends of Harbors, Beaches, and Parks. The group manages the Green Vision Project, a database identifying high-priority conservation parcels covering all Southern California counties except Ventura County. Other projects have been sponsored by local municipalities and land conservancy groups.[38]

Although Orange County had an early start with a lot of political support and landowner buy-in for the NCCP process, the process has not always been easy and free of conflict.

First, consider the case of Crystal Cove State Park, a hotspot for California gnatcatcher sightings year-round (see fig. 3.2). To extend the park's boundaries, the state of California bought a mobile home park called El Morro Village that had been established along the coast—some homes were placed right on the sandy beach—and intended to close the village in 1999. The mobile homeowners pleaded with the state for more time to move their homes away, and so the state set up a lease until 2004, at which time all homes were to be removed. Instead of using those five years to find new land for their homes, many of the residents added elaborate decks and other structures that made the mobile homes immobile and thus required heavy equipment to extract them. The residents then sued the state, claiming that the state was forcing the residents to violate the Endangered Species Act since the heavy equipment would destroy the habitat of several ESA-listed species, including coastal California gnatcatchers and the western snowy plover (*Charadrius alexandrinus nivosus*), which nests on beaches. At oral arguments, the mobile-home residents claimed that "the state's design was to have the Plaintiffs demolish their own homes and, in the course of doing so, kill or scare away any endangered species in the area. By letting the residents kill all the endangered species, the state could undertake its own construction efforts without obtaining requisite take authorization for the plover and without fear of incurring a liability under the ESA. The state hatched this plan, Plaintiffs contend, because it knew it could never undertake the Conversion Project without taking the plover and it knew it could never obtain the requisite take authorization for the plover." Yet the state did have incidental take permits for all listed species, the residents lost the case, and El Morro Village was demolished in 2006.[39]

Next, consider the bizarre case of the El Toro Conservation Area. The area supports 364 hectares of prime coastal sage scrub, forming a vital habitat link between inland protected areas and coastal protected areas in the Orange County Central/Coastal NCCP system. Originally part of the Marine Corps Air Station El Toro that was decommissioned in 1999, in 2001 those hectares were transferred from the Department of Defense to the Federal Aviation Administration (FAA) for a communications tower—the decommissioned air base was to become a commercial airport until public opposition crushed the plan. By 2011, the FAA was looking to offload the property and thought it had a willing owner: the USFWS. The El Toro Conservation Area had been pursued by the USFWS since 1996 for inclusion in the NCCP preserve system due to its high biodiversity and its value as a direct connection between the Coastal and Central reserve networks—the USFWS referred to the potential preserve as the Irvine Wildlife Corridor. The conservation area supported roughly 150 California gnatcatchers, the federally endangered Riverside fairy shrimp (*Streptocephalus woottoni*), a large population of cactus wrens, and several other species of concern, including the orange-throated whiptail and burrowing owl (*Athene cunicularia*). Unfortunately, the USFWS didn't want to take direct ownership of it, fearing that it would also take on potentially budget-busting liability for military waste cleanup. Eager to rid itself of the property, the FAA transferred it instead to the Federal Bureau of Investigation (FBI), which was looking for a shooting range. As of 2020 the area was still owned by the FBI. Note that this conservation failure involved federal landowners only:

private landowners aren't always the primary barrier to establishing protected reserves.[40]

Western Riverside County MSHCP/NCCP plan

Riverside County is big—roughly the size of New Jersey. The county is one of the fastest-growing counties in the state, with growth fueled by residents commuting to work in Orange County from the less expensive housing in Riverside County. The two NCCP planning regions each cover over 400,000 hectares (table 6.1). Although Riverside County had a head start developing HCPs due to the Stephens' kangaroo rat, the county's NCCP plans took over a decade before they were approved: the Western Riverside County MSHCP/NCCP plan was approved in 2004, and the Coachella Valley NCCP plan was approved in 2008. However, the Coachella Valley plan doesn't cover California gnatcatcher habitat, so I won't say more about it. Western Riverside County's NCCP plan encompasses the entire western 5,033 square kilometers of the county; 69 percent of that area supports natural habitats, 15 percent of which is coastal sage scrub. As of 2020, the Western Riverside plan had acquired over 165,000 hectares for its reserve network, some of it co-owned with the Metropolitan Water District and state and federal agencies, meeting 82 percent of its protected-area goal.[41]

Implementation of the Western Riverside County NCCP plan received a significant boost from nongovernmental and quasi-governmental organizations. For example, the Trust for Public Land and the Nature Conservancy have engaged in significant land acquisition and preserve management. The NCCP preserves around the Metropolitan Water District's reservoirs—Lake Mathews, Lake Skinner, Diamond Valley Lake—and the lands purchased as mitigation for the flooded hectares form key portions of the NCCP preserve system. Some of these preserves are monitored by the Western Riverside Regional Conservation Authority (RCA) and the Center for Conservation Biology at the University of California, Riverside.[42]

Overall this NCCP plan was projected to cost a jaw-dropping $1.54 billion over the lifetime of its seventy-five-year permit, with $733.6 million for land acquisition and $805.9 million for land management, monitoring, and contingencies. Like the Orange County Central/Coastal plan, the Western Riverside County NCCP plan uses an endowment fund, with a target size of $100 million in twenty-five years. The county employs a variety of novel funding arrangements for land acquisition, including grants and loans from state and federal agencies, and incentives for landowners and developers including land swaps, density bonuses and transfers, and promises to fast-track project approval. These alternative land acquisition approaches kept the NCCP plan on track when development impact fee revenues dropped during the 2008 economic recession.[43]

A 2017 study by Heather Hulton VanTassel and coauthors found that across western Riverside County, California gnatcatchers lost 11.7 percent of their suitable habitat between 1980 and 2003, before the NCCP plan was approved, and 40.6 percent of their suitable habitat after 2003 when the plan was running. Some of the loss of suitable habitat was due to weather and climate factors, but a substantial portion was lost to development. Much of that post-NCCP habitat loss was on unprotected land; unprotected areas were twice as likely to exhibit a decline in gnatcatcher habitat as protected areas. Although the Western Riverside County NCCP plan has not halted the loss of gnatcatcher habitat, it has made a positive difference.[44]

Once this NCCP plan is complete, 61,917 hectares of private land will be incorporated into preserves. Combined with considerable publicly owned protected areas—114,121 hectares in federal and state ownership, 26,033 hectares owned by local governments—the plan will ultimately create a 202,000-plus hectare preserve system costing $1.5 billion, leaving over 300,000 hectares available for development. Of those 61,917 hectares of private land yet to be incorporated into the preserve system, local municipalities are on the hook to acquire over half of them by any feasible means, and state and federal agencies for the remainder.[45]

When the Western Riverside County NCCP plan was approved, the county did not know which private lands would make up those remaining 61,917 hectares. To manage this uncertainty, the county designated a criteria area of 137,600 hectares from which the 61,917 hectares of the highest conservation value would be identified as development plans were submitted to the county for approval. The RCA was tasked with deciding which areas were needed for the NCCP preserves and which could be released for development. To do this, the authority divided the 137,600-hectare criteria area into 65-hectare cells and assigned each cell a conservation goal and directives for potential development. Changing a cell or a parcel's conservation assessment requires a Criteria Refinement Process, and any land swap within or among cells requires an Equivalency Analysis; both processes allow for public comment.[46]

Some of the land swaps have been complicated and litigious. One case involved the 390-hectare Warm Springs Ranch, owned by Anheuser-Busch Companies, LLC, to raise and train the brewer's famous Clydesdale horses. In 2002, the RCA placed a conservation overlay on the ranch and a year later rezoned the parcel to reduce its allowable development density. These decisions meant that if Anheuser-Busch or any future owners wanted to develop the property, they would be required to go through an additional Habitat Evaluation and Acquisition Negotiation Strategy (HANS) process with the RCA.[47]

In 2005, Anheuser-Busch began the HANS process to develop

the ranch. At the conclusion of the HANS process in 2007, Riverside County informed the company that all but 29 hectares of the ranch would need to be acquired for the NCCP/MSHCP plan. At that point, the company and the county began to negotiate land swaps, which involved transferring ownership of other parcels of low conservation elsewhere to the company for development. But as the land swaps were finalized, the public outcry over the loss of the ranch to development prompted the government to walk away from the agreement. Two years later, when the county tried to use eminent domain to obtain some ranch property for a road project, Anheuser-Busch forced the county back to the negotiating table. Ultimately, all parties agreed to an eight-year, $50 million land purchase schedule for the ranch, in nine phases between 2012 and 2020, since the county couldn't afford to purchase it all at once.[48]

The area of the ranch scheduled to be purchased last, the 81-hectare Phase 9 area, was unique. The agreement for this section required the county to split it off into a separate legal parcel and the RCA to release it from the conservation overlay. If the RCA didn't buy the property by 2020, it would be available for someone else to buy and develop. In the meantime, the RCA sought out other parcels to preserve as mitigation for the Phase 9 property in case it was developed in the future and identified two such properties totaling 431 hectares. But in the process of the Equivalency Analysis for that land swap, biologists discovered two species of concern on the Phase 9 property that were not on the 400-plus hectares of the two swap properties: the California horned lark (*Eremophila alpestris actia*) and Southern California rufous-crowned sparrow (*Aimophila ruficeps canescens*). Therefore, the land swap was not equivalent.[49]

Meanwhile, in reaction to the county lifting the conservation overlay on the Phase 9 area, a local environmental group — the Friends of the Northern San Jacinto Valley, chaired by Albert Paulek — sued to stop the rezoning and threat of development. Ultimately, the courts sided with Paulek, agreeing that the county had to work through the California Environmental Quality Act before making any changes to its original conservation plans. The RCA walked back its overlay decisions regarding Phase 9 and, instead, fast-tracked its acquisition of the ranch as it had planned to do in the beginning, purchasing the Phase 9 property in December 2018.[50]

The moral of this story is that although these land swaps, development density adjustments, and other alternative land acquisition schemes can look very clever on paper, when implemented they can create more headaches than they solve.

San Diego NCCP Plans

Similar to Orange County, the NCCP process in San Diego County was implemented when conservation planning was already in motion, and NCCP plans were developed quickly. For example, when the NCCP Act was passed in 1991, the city of San Diego was already in the middle of a multiple-species planning program required by sewer system upgrades. But unlike Orange County and its few, large landowners, San Diego County is a patchwork quilt of thousands of public and private landowners with smaller ownerships. Compared to the Orange County process, where large landowners, like the Irvine Company, partnered with the county to design and implement NCCP reserves, the negotiations for the San Diego NCCP plans have needed to be more transparent and include more stakeholders, diluting the influence of any one stakeholder over the process or outcome. The transparency is politically necessary to maintain buy-in from the large number of affected landowners, developers, farmers, and even some council members, who frequently argue for private property rights at public meetings. But the dilution of influence has generated mistrust in the process from environmental groups, including the Center for Biological Diversity, Spirit of the Sage Council, and San Diego Audubon, and eventually compelled some cities to break off from the subregional NCCP plan to form their own subarea plans. These splits increase the influence of local governments and landowners over their own land base but create coordination difficulties across the subregion that neither the San Diego Association of Governments nor the Metropolitan Planning Organization can fully resolve.[51]

As a result, San Diego County has six NCCP plans, some of which are further divided into multiple subarea plans (table 6.1) — most are still in the planning stages. Two of the plans, the San Diego County Water Authority and the San Diego Gas and Electric Subregional NCCP plans, concern linear features, such as utility lines, and overlap with other NCCP plans, similar to the situation for the Orange County Transportation Authority NCCP plan. Thus, their plans cover every species of concern their projects could encounter and include protected areas that are not protected by the utilities themselves.

Four other plans are NCCP plans in the traditional sense (pay close attention to the subtle difference in acronyms that distinguish these plans). The San Diego East County Multiple Species Conservation Program (MSCP) is mainly too far east of the California gnatcatcher's range, so I won't discuss it further. The other three NCCP plans cover coastal sage scrub with the highest densities of California gnatcatchers in the region: the San Diego North County MSCP in the northwest corner of the county; the San Diego County MHCP in the urbanized southwestern corner of the county; and the San Diego South County MSCP, sandwiched between the San Diego East County MSCP and the San Diego County MHCP. As of 2014, San Diego County had 89 publicly owned reserves larger than 100 hectares enrolled in NCCP plans, and NCCP-enrolled areas made up almost 81,000 hectares of all habitat types.[52]

Approved in 1998, the San Diego MSCP/NCCP plan covers an area twice as large as the Orange County Central/Coastal NCCP plan, with four times the coastal sage scrub but only half of it protected. Protecting the remaining half was estimated to cost $262–$360 million at the time of the plan's approval. The plan covers all habitats and species of concern, not just coastal sage scrub species and the coastal California gnatcatcher, although the gnatcatcher features prominently in many conservation efforts. This MSCP/NCCP plan benefited from early adoption of the NCCP process by local governments, with four subarea plans for the city of Poway (1996), city of San Diego (1997), county of San Diego (South County, 1998), and city of La Mesa (1999) approved before 2000. The MSCP/NCCP plan covers eighty-five species and 234,718 hectares of the far western portion of the county; the city of San Diego represents about half of those hectares. The plan is to set aside 69,574 hectares in preserves, including 21,044 hectares within the city of San Diego, funded by a mix of federal, state, and local funding. The San Diego MSCP plan monitors target species, degree of habitat fragmentation, and usage of wildlife corridors to measure progress toward conservation goals.[53]

Along with city officials, the USFWS made some controversial decisions in the permitting of the San Diego MSCP plan that led to lawsuits. In one instance, the Service approved a permit to destroy more than sixty vernal pools for one commercial development project, the Cousins Market Center in Mira Mesa. In response, fourteen environmental groups sued the USFWS over the incidental take permit process associated with the San Diego MSCP plan, and the U.S. district court agreed to nullify the NCCP plan's coverage of vernal pools. In that case, the court cited the fund-as-we-go mechanism for the mitigation process as inappropriate for all species covered by the NCCP plan, not just vernal pools. The court case renewed the county's focus on funding mechanisms for the plans.[54]

Rancho Palos Verdes (Los Angeles County)

Wikipedia suggests that Spanish ranchers gave Rancho Palos Verdes—translated as Ranch of the Green Sticks—its moniker due to an impressive stand of willows on the Palos Verdes Peninsula. *Palo* can also refer to sports equipment such as hockey sticks and golf clubs; this latter definition seems more appropriate now for the 52,000-hectare area with six golf courses. Although the Palos Verdes Peninsula was originally owned by the Sepulveda and Bixby families for cattle ranching and farming, its desirable coastal views drove its early residential development. Already in 1913, the area was promoted by developer Frank Vanderlip as the "most fashionable and exclusive residential colony" nationwide. Residential development began in earnest in 1922, using plans designed by Frederick Law Olmsted Jr., and three cities were incorporated between 1939 (Palos Verdes Estates) and 1957 (Rolling Hills and Rolling Hills Estates). Sensing the impending rush to develop, residents in the surrounding areas incorporated as the city of Rancho Palos Verdes in 1973, prioritizing low-density land use and low taxes in its charter.[55]

The Rancho Palos Verdes NCCP/HCP planning region encompasses a 3,487-hectare coastal area, including 513 hectares of coastal sage scrub supporting 26–56 pairs of California gnatcatchers. The habitat preserve network consists of 568 hectares of public areas, 8 hectares owned by the Palos Verdes Peninsula Land Conservancy that serves as the habitat manager for the NCCP/HCP plan, and 69 hectares owned by a mixture of public and private owners. Enrollment agreements were signed in 1992 with the city of Rancho Palos Verdes and the city of Rolling Hills, and the planning agreement was

Box 6.1. White Point Preserve, Palos Verdes
Peninsula, Los Angeles County, January 15, 2017

My sister—an actor working the usual two jobs in Los Angeles—finally relented on a rare day off and accompanied my son
and me to see the California gnatcatchers I had been rambling
on about for over twenty years. We joined a crowd of joggers
and dog walkers on the short yet scenic trails. Although I had
always known about this small but persistent population of
gnatcatchers on the Palos Verdes Peninsula, I had never made
the trip to see them. We spotted a pair almost immediately; in
all, I found at least three pairs in this small chunk of coastal
sage scrub hugging the ocean. The high density of California
gnatcatchers persisting on such a small patch provides hope that
these pocket parks, even with a lot of visitors, can play an active
role in urban species conservation.

signed soon after in 1996. It took another eight years to get implementation agreements signed and to complete an Environmental Impact
Report. Plan documents were finalized in 2018 but had not yet gained
approval as of May 2020.[56]

The Rancho Palos Verdes NCCP/HCP plan covers six plants
that are state species of concern, two federally endangered butterflies—the Palos Verdes blue (*Glaucopsyche lygdamus palosverdesensis*)
and El Segundo blue (*Euphilotes battoides allyni*)—and the California
gnatcatcher and the cactus wren. Like the Western Riverside County
NCCP plan, the Rancho Palos Verdes plan benefits from governmental funding; a $607,250 grant from the Wildlife Conservation Board
in 2013 allowed the city to purchase 23 hectares of coastal sage scrub
for the preserve network. The NCCP plan is not the only habitat con-

servation policy operating in the county, but it is one of a few that focus on strict protection of habitat, not multiple uses. For example, Los Angeles County identifies Significant Ecological Areas in planning documents, but these are not strict preserves and permit such land uses as oil and gas pumping.[57]

Of all of the NCCP plans, the Rancho Palos Verdes NCCP plan is one of the most isolated in Southern California. California gnatcatchers and other coastal sage scrub species need to traverse a sea of residential neighborhoods, shopping districts, and freeways to seek new habitat and meet others of their kind. The isolation of this NCCP plan emphasizes the need for protected areas that could serve as connective stepping-stones, corridors, and matrix greening (box 6.1; see also chapter 8).

7

Is the NCCP Policy a Success?

Thirty years ago, the Natural Community Conservation Planning policy began a regional experiment in collaborative species conservation. Assuming that three decades is enough time to evaluate a conservation policy—and it might not be—what can be said about the NCCP approach? Has it set aside enough habitat for the coastal California gnatcatcher to ensure its long-term existence in a rapidly urbanizing region? Has it prevented more Endangered Species Act listings? If it was successful, why? If the NCCP policy is promoted as a successful strategy to prevent further biodiversity declines, we must know the answer to these questions.

Ecological Success

Success of the NCCP policy can be narrowly defined as stable coastal California gnatcatcher populations and adequate protection of its habitat. If at any point the NCCP policy fails to stabilize gnatcatcher populations, then the 4(d) special rule can be revoked, returning the coastal California gnatcatcher to the U.S. Fish and Wildlife Service–governed ESA process. By these gnatcatcher-focused measures, the NCCP policy has been fairly successful. As of 2020, thousands of hectares of coastal sage scrub were protected or were slated to be, with protection running for the next fifty to seventy-five years (see table 6.1). The California gnatcatcher has had relatively stable population numbers, we think—it's hard to say without adequate, regionwide monitoring data. According to their genetics, California gnatcatcher populations in the United States are well connected, with only the populations in Ventura County exhibiting genetic isolation, although the populations in Palos Verdes, San Dimas, and the Coyote and San Joaquin Foothills appear somewhat isolated as well. Stable demographics plus genetically connected subpopulations suggest that the current conservation and management strategies are working.[1]

But the NCCP policy was meant to protect all coastal sage scrub species, including two other target species: the coastal cactus wren and orange-throated whiptail. How have these species fared? Cactus wrens are in decline and displaying evidence of genetic isolation, and the status of orange-throated whiptails is still unknown due to a lack of information. While none of the hundreds of species covered in approved NCCP plans have been listed under the federal ESA since the NCCP policy began, the lack of new listings could be due to insufficient data and politics, not biology.[2]

Some nontarget species have fared better, such as the Quino checkerspot butterfly and the Pacific pocket mouse (*Perognathus longimembris pacificus*). Quino checkerspot butterflies have increased and are already demonstrating an ability to adapt to climate change using range shifting. The appearance of Quino checkerspot butterflies in the San Diego MSCP planning area triggered a separate Habitat Conservation Plan for them, since it was not one of the species covered in the approved NCCP plan. The Pacific pocket mouse was presumed extinct for twenty years until a small colony of thirty-nine mice was found at Dana Point in Orange County in 1993, prompting the USFWS to issue an emergency endangered listing for the species. Since then, several small populations have been found along the coast from San Diego to Ventura Counties, some in coastal sage scrub habitat. Like the Quino checkerspot butterfly, individual HCPs are necessary because the species was not covered in NCCP plans.[3]

What has happened in areas without NCCP plans? Ventura County was excluded from the coastal sage scrub NCCP pilot program, and the county has faced a growing list of federal- and state-listed species (or candidates for listing) that have delayed development projects, including the California red-legged frog (*Rana draytonii*), San Fernando Valley spineflower (*Chorizanthe parryi* var. *fernandina*), Belding's savannah sparrow (*Passerculus sandwichensis beldingi*), and least Bell's vireo. As climate change drives species and ecosystems to cooler, higher, more northerly ground, the exclusion of Ventura County from the pilot NCCP process may be one of the policy's largest oversights. Ventura County and San Bernardino County have been slow to engage with the NCCP process generally, and I suspect that will be problematic for the coastal California gnatcatcher and other coastal sage scrub species going forward. There are certainly predictable California gnatcatcher populations in both of these counties that merit protection (box 7.1; see also box 3.1).[4]

Detractors of the NCCP approach lambast the use of poorly chosen surrogate species, inadequate data, and underfunded management activities, although these are common criticisms of Multiple Species HCPs and single-species HCPs as well. Across the country, MSHCPs have mixed results for biodiversity protection and recovery, often delivering worse conservation outcomes than single-species HCPs. Much of this failure is due to the need or desire to start the planning process before adequate data on all species and habitats are available, placing too much weight on the use of surrogate species that may turn out to be poor indicators for the species community as a whole. MSHCPs and NCCP plans encourage the inclusion of all potentially present species in a plan, although whether

---

Box 7.1. Open space east of Redlands Municipal
Airport, San Bernardino County, March 12, 2017

California gnatcatchers had been reported here fairly consistently on eBird, but I had my doubts. In a satellite image the habitat looked like a small patch of coastal sage scrub with a strip of riparian running through it, surrounded by the airport, a rock quarry operation, orange groves, and housing developments. Even my son was skeptical. "This is not gnatcatcher habitat," he said, as we stared at a fenced orange grove on one side of the street and a high school on the other. Since we had driven a long way and it was considered a birding hotspot by locals, I decided that we'd walk down the road a bit and look around. We didn't have to walk much past the rock quarry to find coastal sage scrub at the end of the airport runway. As I pished at a California towhee perched on a bush to get it to turn around, a male gnatcatcher popped up into the top of the bush right next to it. He looked around silently, then dropped back down. I pished again and managed to rustle up the female, but only briefly; clearly, they had an active nest. Finding this pair was a stroke of luck brought on by that towhee. On the drive home, I wondered how well that spot was protected from development simply due to its proximity to the airport.

---

those species are actually present in the finalized preserve networks is rarely confirmed. For example, in a 2006 survey of twenty-two MSHCPs in USFWS Region 1 including Southern California, 41 percent of covered species had not been verified as present in the planning area. Although MSHCPs are supposed to give conservation outcomes and regulatory certainty equal weight, in practice the certainty of development today is prioritized over long-term conservation needs.[5]

Furthermore, although species protection is the goal, incidental take permits in MSHCPs and NCCP plans are commonly habitat based because habitat is far easier to measure and monitor. This approach risks underestimating the anticipated take of individuals of listed species. As explained in chapter 6, this happened to the NCCP policy early on. By 1996 NCCP plans for coastal sage scrub collectively had issued more incidental take permits for coastal California gnatcatchers than the 5 percent cap on gnatcatcher pairs allowed, even though the program remained under the 5 percent habitat cap. This risk of inadequate protection is particularly problematic for rare species, for which ecologists often have less information — these species are poorly served when bundled into an MSHCP.[6]

Adaptive management, as called for in the NCCP Act, is nearly impossible without adequate information on the results of management actions or changing conditions inside and outside the preserves. All HCPs, MSHCPs, and NCCP plans are required to implement monitoring plans to assure that biodiversity is being successfully protected, although this requirement is rarely met. Monitoring plans should be designed to trigger actions when populations drop below a threshold, such as a moratorium on issuing additional incidental take permits, since monitoring without action does little good. For example, plans rarely address the increased risk to protected species from human uses of preserves, even though some types of human access can significantly degrade a preserve (such as Carlsbad Highlands, chapter 6). In near-city reserves, protected species can be exposed to high levels of recreation due to the popularity of these open spaces. On the bright side, NCCP plans, MSHCPs, and HCPs, if well funded, do provide excellent opportunities for basic and applied research in ecology, conservation biology, and genetics.[7]

Gathering necessary information and executing management actions can be extremely expensive. Environmental groups and scientists correctly criticize the approval of NCCP plans and MSHCPs without secure funding mechanisms. Although a variety of funding tools are possible, many plans rely on development fees, taxes, or bonds simply because these are more straightforward for the public entities that administer the plans. Although they were meant to be an innovative funding scheme for NCCP plans, conservation banks and other market-based funding mechanisms are more commonly used by private entities or quasi-governmental agencies in their MSHCPs, particularly for large energy or water infrastructure projects.[8]

Even though multispecies plans can offer some economies of scale over HCPs, adequate funding for conservation activities is a difficult obstacle to surmount. Resource deficiencies were a regular occurrence during the first ten years of NCCP implementation. In the long term, NCCP plans are more successful if they estimate resource needs and identify ways to meet them early on than if the plans proceed in ignorance. For example, the Orange County Central/Coastal NCCP plan and San Diego MSCP/NCCP plan estimated that the combined management and monitoring costs would be roughly $16 per hectare for each plan at the time they were proposed, which is about $27 per hectare in 2020 dollars. As a point of reference, management costs of the approximately 1,000-hectare San Bruno Mountain HCP, which covers just two butterfly species in an urban park with a lot of visitors, has been twice as expensive to manage and monitor, clocking in at around $33 per hectare in 2001—about $48 per hectare in 2020 dollars. In 2001, the San Diego Association of Governments estimated that fulfilling all the planned land acquisition and management actions for local municipalities in the San Diego MSCP/NCCP plan would cost $1.26 billion, or $1.8 billion in 2020 dollars. Considering that these municipalities must get two-thirds of voters to approve all revenue-raising programs, these levels of funding needs are daunting. Luckily, funding for nature preservation in California has received considerable support from private foundations and nongovernmental organizations, such as the Trust for Public Land, the Nature Conservancy, and the David and Lucile Packard Foundation. The Packard Foundation's Conserving California Landscapes Initiative donated $175 million to support land acquisition, conservation easements, and conservation programs on 138,402 hectares in northern and central California, attracting almost $1 billion in matching funds from other donors.[9]

Political Success

If we are curious to know whether the NCCP model might work in other states, we must take a broader view. The species and ecosystems are likely to be different, and many of the threats might be different, too. On the other hand, the politics are likely to be quite similar, including a familiar mix of conflict between conservation and land use on private property and local governments with inadequate budgets to acquire the necessary land. Indeed, through its procedures for formalizing privately owned conservation areas into monitoring plans, the NCCP policy could address the transparency gap in the effectiveness of conservation easements and other tax incentives for conservation on private land. After thirty years, we can compare the NCCP policy's success across the different counties and identify several characteristics of the political context that boost the likelihood of an NCCP plan's success: amenable land-use traditions, active third parties providing oversight, strong stakeholder coalitions, and reduced landowner uncertainty.[10]

One of the strongest predictors of NCCP plan success is amenable land-use planning history and tradition. The two early NCCP policy adopters, Orange and San Diego Counties, were engaged in forward-thinking, integrative land-use planning years before the NCCP policy began. Orange County has historically had a strong land-use planning tradition among local municipalities and developers, as demonstrated by the Irvine Company's emphasis on planned communities. The city of San Diego was already planning a habitat preserve network with its sewer infrastructure upgrades. The NCCP policy simply incentivized land-use planning approaches that local governments were using. Ventura County also has a history of proactive growth management and planning activity, made actionable by policies to corral urban growth near city centers and protect agricultural land. Should Ventura County enter the NCCP process, I would expect it to experience success similar to that seen in Orange County.[11]

In contrast, even though Riverside County was already grappling with the Stephens' kangaroo rat when the NCCP process began, the county was slow to move. Planning in Riverside County tends to be more passive, with significant deference to the building industry. Los Angeles and San Bernardino Counties have lagged even further behind in large-scale, integrated planning. In Los Angeles County, planning is heavily influenced by politics and finances, and only the Rancho Palos Verdes NCCP plan has moved forward. This is bad news for viably connecting California gnatcatcher populations in Ventura County to those farther south and east. Planning in San Bernardino has historically been firmly progrowth, anticonservation, and occasionally problematic—a recent public scandal erupted over alleged corruption involving development permits in the city of Upland. San Bernardino's recalcitrance toward conservation policy is problematic for the California gnatcatcher and other coastal sage scrub species. The valuable stretch of the Santa Ana River basin in San Bernardino County serves as a major wildlife connection between the San Gabriel and Santa Ana mountain ranges and is home to ten federally listed species. The Upper Santa Ana River HCP—not affiliated with an NCCP plan—finally reached the public comment stage of the approval process in December 2019. The HCP applicants, the San Bernardino Valley Water Conservation District and the San Bernardino County Flood Control District, sought incidental take permits for four listed species, including the coastal California gnatcatcher and the San Bernardino kangaroo rat (*Dipodomys merriami parvus*).[12]

The influence of independent organizations should not be ignored either. In successful NCCP plans, third parties, such as residential associations and environmental interest groups, gathered data, highlighted gaps in plans and coverage, and kept participants honestly engaged with the plan. The Safari Highlands Ranch development project in San Diego County is a perfect example of the influence of third parties in the NCCP process. Encompassing 444 hilly hectares of coastal sage scrub, oak woodlands, and riparian forests, Safari Highlands Ranch is situated just north of the San Diego Wild Animal Park. The area supports five to ten pairs of California gnatcatchers as well as other species of concern, and it was included in the county of San Diego NCCP/MSCP planning area, straddling the boundary between two subregional MSCPs. Under county zoning, the area was classified as rural land use, which limited development density to 27 homes in total on the parcel. But the parcel was within annexing distance of the city of Escondido, which had started the NCCP process in 2001 but made so little progress on its plan that the USFWS refused to extend the 4(d) special rule to the city. If the project area was annexed by Escondido, the area could be rezoned as residential, allowing for 550 homes. That move in jurisdiction would not only have increased home density allowances but would also have removed the land from the NCCP process and streamlined development, or so the developer may have thought.[13]

The residents living next to the area were not thrilled by the prospect of additional traffic, wildfire risk, and changes to the aesthetics of their rural landscape. Their association, the San Pasqual Valley Preservation Alliance, challenged the project with the support of a variety of social and environmental groups, and the San Pasqual Band of Mission Indians. The alliance also publicized the concerns with the project of the San Pasqual Union School District, the USFWS, and the California Department of Fish and Wildlife. Although the developer changed the name of the development project to Harvest Hills to try to move the project forward, as of May 2020 the project was stalled at the planning stage.[14]

Since the NCCP policy is dependent on groups of cooperating stakeholders to assemble and manage habitat reserves, one might think that Southern California is the worst place to try such a policy. The region is cursed with a legacy of local, narrow, disconnected policies, ignorant of their impacts at larger scales. Indeed, the lack of regional planning and cooperation is a time-honored tradition in California, as touched on in chapter 1, and is exemplified by decades of failed open space planning. In 1930, the urban planning firms of the Olmsted Brothers and Harland Bartholomew and Associates generated a master plan for parks and recreation throughout the Los Angeles region, including an open-space corridor along the Los Angeles River. The plan's authors emphasized that the rapid urbanization of the region threatened to destroy the natural amenities that attracted the influx of newcomers in the first place. The Los Angeles Chamber of Commerce, which had commissioned the plan, was horrified by the potential threat the plan posed to private property rights and promptly shelved the report. Ninety years later, the city is finally restoring the Los Angeles River to stabilize water supplies, reduce flooding risk, and function as an open space corridor.[15]

Given that the NCCP approach depends on collaboration among many stakeholders and jurisdictions, the fact that it has had any success at all in California's hyper-fragmented government is a good sign. At the time of a programmatic review of the NCCP policy in 2001, environmental organizations were divided in their support, even within affiliated groups—San Diego Audubon was opposed to the policy, while California Audubon supported it. Local governments and developers doubted the regulatory certainty that it promised.[16]

Two decades later, the success of its coalition-building approach is clearer. But the strength of these coalitions is profoundly influenced by sociopolitical context. Of the six California counties in the gnat-catcher's range, Orange County has implemented the NCCP most successfully. Its achievement is due in part to key buy-in from the Irvine Company, which owns a significant amount of land, and in part to the momentum the county gained from its experience with HCPs before the start of the NCCP policy. The NCCP process in San Diego County, where no large landowners could adequately champion the policy, has been slower and has more obviously suffered from Southern California's political habit of fragmentation and decentralization.[17]

Finally, the NCCP policy has streamlined the regulatory process for species conservation and improved certainty for developers and planners. The policy spreads risk and uncertainty from landowners out to many other stakeholders, including back to the federal government through the 4(d) special rule and the No Surprises guarantee. The NCCP policy also grants state and local agencies increased discretion and adaptive capacity to adjust to new conditions or unforeseen effects, within limits. As intended, local governments have used the NCCP policy as an incentive for planning in a proactive and systematic way, reducing environmental and political risk. Some would argue that the NCCP policy reduces landowner risk and uncertainty too much, given the lack of adequate scientific data and secure funding schemes to ensure that NCCP plans preserve the species and habitats they set out to preserve. The state's Legislative Analyst's Office has repeatedly issued warnings over the lack of monitoring and evaluation of the NCCP pilot program in Southern California, specifically the lack of measurable goals and objectives. Neglecting the policy evaluation stage of the NCCP process is a severe shortcoming, increasing the probability that more species will need to be listed by the ESA in the future.[18]

Environmental groups serve an important corrective role here; they have the ability to sue if plans fail to protect listed species. The USFWS and the California Department of Fish and Wildlife have the authority to revoke permits only if NCCP plan participants violate the plan's implementation agreement, not when the plans have been followed but have failed. Once an NCCP plan is approved, no additional mitigation requirements are allowed under either the federal or state ESA, if even the NCCP turns out to be insufficient for preventing the extinction of the California gnatcatcher or other listed species due to unforeseen circumstances (chapter 5). The only failsafe that the USFWS has is an extraordinary circumstances clause, which allows the USFWS to renege on the No Surprises and other guarantees. Given that some of these NCCP permits span seventy-five years, the lack of ongoing programmatic evaluation increases the odds that species will be listed under the ESA when extraordinary circumstances occur. Across the board, uncertainty of future conditions has been poorly accounted for in permits or plans. The No Surprises guarantee and incidental take permit conditions do not account for greater uncertainty for species at high risk of extinction or for those with insufficient scientific data. Thus, while perceived uncertainty may have been reduced, real uncertainty lingers in the distance. For this reason, I predict an increase in petitions and lawsuits from environmental interest groups to list species and force additional protection due to extraordinary circumstances.[19]

Assessing success or failure might not even be possible for a policy as complex as the NCCP policy. Instead, assessing a range of achievement in different dimensions might be more appropriate. Using Allan McConnell's policy success spectrum from "success, resilient success, conflicted success, precarious success, to failure," we can examine the NCCP policy's progress in three critical dimensions: *process, programs,* and *politics.* The NCCP *process* involves decision-making via large coalitions of stakeholders, maintaining coordination among municipalities through oversight committees, and identifying the conservation value of land and prioritizing it for preserves. NCCP *programs* include the mechanisms for enrolling sufficient property, such as regulatory incentives and land swaps, and the implementation of regionwide surveys and their use in adaptive management. Finally, the *political* success of the NCCP policy is measured in its acceptability to a wide variety of stakeholders and interest groups and in the enthusiasm with which elected and appointed officials promote and support it.[20]

In this framework, the NCCP has achieved conflicted success in its process and programs, given the lack of coordination among NCCP plans, insufficient data and funding resources, slow accumulation of land into preserves, and inconsistent oversight by overworked USFWS and California Department of Fish and Wildlife staff. Politically, the NCCP has been a resilient success with some opposition—especially among environmental groups. Most stakeholders are supportive, including the USFWS, and plans are moving forward in the approval process. We even have some indication that the NCCP policy's political success has increased over time. In 2001, an Orange County planner pessimistically commented that "[the NCCP is] not an abject failure, but close to it." In 2019, few remained this cynical. Although the NCCP policy has not yet had complete success, it has certainly moved away from the realm of abject failure.[21]

The Future of the NCCP Policy

Despite some concerns, the California gnatcatcher continues to be a flagship or surrogate species for the NCCP policy in Southern California. Although there are many unknowns about the gnatcatcher—accurate population counts and trends in the United States and Mexico, shifting ranges in response to climate adaptation—we know far more about the California gnatcatcher than about most other species in the region. As a surrogate species, the California gnatcatcher has a lot of benefits. The gnatcatcher's short life cycle and high fecundity make it responsive to changes from the local to the landscape scale, and its conspicuous and year-round presence makes it relatively easy to track. With the decline and isolation of cactus wren populations, it may be prudent to shift conservation focus toward that species and others that are more sensitive to coastal sage scrub fragmentation. As NCCP plans identify new surrogate species and develop outside the California gnatcatcher's range, the NCCP policy will decouple from the fate of the gnatcatcher. As this happens, the future success of NCCP plans will face four challenges: decommissioning of large military bases, climate change, ongoing human population growth, and protecting evolutionary processes in cities.

Decommissioning the region's large military bases may generate problems for adjacent NCCP plans. Although these bases represent opportunities for the federal government to convert them into wildlife refuges, in practice this is difficult to do—recall the problematic transfer of the Marine Corps Air Station El Toro discussed in chapter 6. The habitats on large military installations, such as Camp Pendleton, are already factored into the NCCP plans in San Diego and Orange Counties as breeding habitat for core populations, under the assumption that these habitats will be protected in perpetuity. But computer simulations of the interaction among fire, coastal sage scrub, and grassland on Camp Pendleton suggest that the California gnatcatcher and Pacific pocket mouse may face an uncertain future—perhaps more so than cactus wrens and Stephens' kangaroo rats, even though the long-term persistence of these latter two species is uncertain, too. If these military areas are decommissioned and developed, the NCCP policy's success is doubtful; these areas offer a significant margin of error for conservation under the specter of climate change impacts.[22]

Although climate change is one of the NCCP policy's biggest challenges, climate impacts are poorly incorporated into existing plans. Conservation areas contribute to California's climate change mitigation strategy through preservation activities that protect existing carbon sinks and restoration activities that build them. Despite this, the 2008 Desert Renewable Energy Conservation NCCP plan is the only plan that addresses climate change directly. It wasn't until 2009 that the California Natural Resources Agency prioritized climate change adaptation in biodiversity and habitat conservation plans—too late to influence the NCCP pilot program for coastal sage scrub. The cumulative amount of protected and soon-to-be-protected land under all the approved NCCP plans and HCPs in the region should protect most native plant species into midcentury. Whether the protected habitat is configured to help species move around the landscape is less clear. There is a risk that NCCP preserves may leave some species, such as the gnatcatcher, insufficiently protected in the future if climate change pushes gnatcatchers into areas where habitat has been destroyed by urbanization.[23]

Already the most populous and urban state in the United States, California is expected to be home to fifty million people by 2030; four million already live within the city limits of Los Angeles. The Los Angeles–San Diego megalopolis has expanded eastward into Riverside–San Bernardino to become a megaregion, and its core—Los Angeles–Long Beach–Anaheim—is the densest urban area in the

United States with roughly eighteen thousand people per square kilometer. The megaregion is bounded physically by the Pacific Ocean and a ring of mountain ranges and politically by the U.S.-Mexico border, which constrains where newcomers can be housed. Affordable housing shortages plus a growing population create a massive pressure to build. Even so, the increase in land converted to houses, stores, roads, and other gray infrastructure far exceeds population growth, so that each 1 percent increase in people results in a 3 percent loss of natural areas. That the NCCP policy has managed to protect over 400,000 hectares of habitat statewide under these conditions should perhaps be viewed as heroic.[24]

Finally, adaptation and evolution happen in cities at the individual and community levels. Urban areas generate a host of unique pressures and disturbances on species: new surfaces and structures, novel chemicals, and light pollutants. These urban conditions drive species' adaptations, such as longer legs and stickier feet, which can better scale metal and glass, in urban crested anole lizards in Puerto Rico. Dozens of new species of evening primrose (*Oenothera* agg.) have evolved in European cities after individual plants were brought back from North America by European explorers 350 years ago. Evolution happens at the level of communities, too, as novel combinations of species form new habitat associations and ecosystems, which ecologists are just beginning to explore. That said, the conservation value of novel ecosystems is incredibly controversial. Urban communities represent a globally ubiquitous collection of species—the small subset of species that can survive the urban filter. There is little evidence that novel species and communities can offer a reasonable proximity in structure and function to the species and communities that are replaced. How these novel communities will interact with the denizens of coastal sage scrub is anyone's guess.[25]

The Next Generation of Conservation Policy

By this point in the book I hope that you are convinced that relying on the ESA as our sole conservation tool is a bad idea and that additional policies such as the NCCP policy are necessary. The ESA is triggered when a species is circling the extinction drain, when it is much harder and more expensive to conserve. The ESA is really thus a last resort. Its single-species focus also provides fodder for endless litigation over species designations, an artifact of eighteenth-century thinking that isn't appropriate or feasible for most taxonomic groups. Furthermore, the ESA is mute on whether there should be no net loss of habitat; it allows habitat loss up until the point at which additional loss would place the listed species in jeopardy. Without habitat restoration, this ultimately results in less habitat and more imperiled species. An attempt initiated by President Barack Obama in 2015 to expand the USFWS's policies toward habitat mitigation and net habitat gain was undone three years later, due to concerns over the legality of the net gain target.[26]

What are our other options?

Shifting our conservation focus to habitats is the obvious first choice. Even now, the ESA's focus can be extended to habitats via the critical habitat designation and Section 10(a) permitting phases, mainly through developing MSHCPs. As I have discussed, MSHCPs are quite popular but have a dubious track record for conserving all of the species covered under them. With its primary focus on habitats, the NCCP approach may be a more nuanced and politically savvy approach than what the ESA offers. But habitats can be tricky to delineate and are constantly moving targets. Habitats change over time due to natural forces, such as ecological succession, and disturbances, such as fires and floods. A focus on habitat will also not move us past litigation over definitions, as illustrated by the 2018 Supreme

Court case over whether designated critical habitat requires that it be habitable.[27]

Ecologists are working on other conservation targets beyond species diversity, which could alleviate some of these issues. Functional diversity classifies organisms by the ecological functions they perform, such as sequestering carbon, cycling nutrients, or stabilizing ecosystems through predation. This kind of diversity would directly preserve the valuable ecological roles that species provide and might be easier to restore, particularly in urban areas where novel species and communities are dominant. Evolutionary hotspots are another possible conservation target. These areas spawn a large number of species or genetically unique populations and may be important in the future for species' adaptations to climate change and urban influences.[28]

Local governments continue to experiment with large-scale, integrated planning, beyond the NCCP approach and other biodiversity conservation strategies. For example, SANDAG is experimenting with land banking to mitigate transportation infrastructure projects, purchasing ecologically valuable land when its price is low and placing it in reserve. Western Riverside County is developing a Comprehensive Integrated Plan that combines NCCP/MSHCP efforts with transportation infrastructure and housing developments. At the state level, the Regional Advance Mitigation Planning (RAMP) program and the Regional Conservation Investment Strategies program, both managed by the California Department of Fish and Wildlife, are similar to the NCCP policy but focus on mitigation and restoration of all native habitat types and ecosystem services. The RAMP program has proved particularly adept at minimizing the negative environmental effects of gray infrastructure projects.[29]

The statewide California Habitat Conservation Planning Coalition, formed in 2009, generates funding opportunities for land acquisition, preserve management, and other necessary NCCP and HCP actions. These opportunities include bonds and revenue from carbon cap and trade auctions via the state's Global Warming Solutions Act (AB 32). In 2015, the coalition scaled up its efforts to form a National HCP Coalition, meeting annually to disseminate information and to advocate for increased funding for the Cooperative Endangered Species Conservation Fund (administered by the USFWS for HCP planning assistance and land acquisition). It is doubtful that any of these landscape-scale land planning policies would be running if their precursor, the NCCP policy, had been a political failure.[30]

From Nature Beyond to Nature Within

Southern California may have unique biodiversity, but the challenges to protect it are not. Throughout the world, as cities swell to absorb millions more humans, urban regions will have to contend with species imperilment, climate change, flooding from impervious areas, invasive species, and novel ecosystems. As the most populous, biodiverse, and fastest growing state in the nation, California can provide valuable lessons to reconcile urban growth with biodiversity conservation and contribute to a broader understanding of conservation in urbanized regions.[31]

Until now, protection of urban biodiversity has occurred outside of cities; this is by design. Many policies and funding mechanisms seek to concentrate development near urban centers and place preserves and open space in the periphery where land prices are lower. Although that approach protects vulnerable species from urban activity, it sacrifices biodiversity in the urban core and reduces the value of the preserves to urban residents who can't afford to travel out to enjoy them. Much of the 15 percent of global land area that has been protected for biodiversity is established in areas of few human settlements and low biodiversity: mountains, deserts, or high-latitude areas. In Southern California, the vast majority of protected areas are national forests in the mountains and national parks in the deserts (see fig. 5.1). While these areas bring the state close to the international Aichi Biodiversity Target of protecting 17 percent of all terrestrial habitats, scientists are now calling for a preservation target of up to half of all habitats—a Nature Needs Half campaign target that Southern California's development has made all but impossible, even with the NCCP policy. Ironically, the movement of rural populations to cities has made this 50 percent target more feasible for all regions except urbanizing ones.[32]

The conservation conundrum is the same everywhere. Static protected areas far from cities are necessary but not sufficient for long-term biodiversity preservation, particularly as urban sprawl decays the remoteness of these areas. Our larger and increasingly urban population is also becoming more affluent, amplifying demand for food that increases the amount of land dedicated to agriculture. As cities sprawl into agricultural areas, as happened in Southern California in the early twentieth century, farmlands are pushed into natural areas. Migratory species and large carnivores that spend much of their time outside of the small protected areas in the periphery are left wholly unprotected, as they use vast swaths of land and sea that are too large to set aside from human activity.

What can be done? We need to learn how to protect species within our cities.

8

Concrete Jungles and Granite Gardens

The city is a granite garden, composed of many smaller gardens, set in a garden world. Parts of the granite garden are cultivated intensively, but the greater part is unrecognized and neglected.

Anne Whiston Spirn, *The Granite Garden*

Cities tend to be established in productive areas, where water, forests, and other ecosystems offer an abundance of natural resources that urban residents need. Because biodiversity is higher in productive areas, urbanization disproportionately affects areas of high biodiversity. Given how dramatically urbanization changes the landscape, many cities are hotspots of threatened and endangered species. Globally, 8 percent of the International Union for Conservation of Nature's Red List terrestrial animals are at risk of extinction due to urbanization, as are half of the ESA-listed and imperiled species in the United States. Urbanization threatens over half of the federally endangered species in California, Florida, and Texas. But cities can act as a refuge for species in peril as well. In Australia, 30 percent of protected species occur in cities, often representing unique communities found nowhere else. In some cases, endangered species have become dependent on the stable resources provided in cities; endangered swift parrots (*Lathamus discolor*) using cultivated fruit and flowering trees in Australian cities are one example. But these at-risk species using

Box 8.1. Claremont Colleges Bernard Field Station,
Los Angeles County, May 18, 2017

One block away from our house, the Claremont Colleges' Bernard Field Station was burning. We were first alerted to the fire by the swarm of news helicopters hovering over our backyard as three county fire department helicopters whirred past them to drop water on the fire. As my son and I walked down Foothill Boulevard toward the smoke, I realized that the fire was burning two of the most active study areas on the station: pHake Lake and the Hill. I was literally watching field equipment go up in smoke, including the scrub jay feeders belonging to Professor Rachel Levin, my former instructor at Pomona College and now good friend. Such are the perils of ecological research in fire-prone ecosystems. With any luck, in a few years the sage scrub that recovers from the ashes could make a fine home for a pair of California gnatcatchers.

urban areas can be further threatened by the abundance of exotic, ornamental species that humans introduce into cities, some of which become invasive. In established urban parks and suburbs, ornamental and invasive species can inflate diversity for a time, until native species are driven away. The complexity of species diversity and its protection in cities signals that we must integrate conservation efforts within cities with the efforts occurring in the region beyond them.[1]

So far, this book has focused on the habitat preserves around the periphery of Southern California's cities. Natural Community Conservation Planning preserves safeguard native species by providing them refuge: sheltered breeding areas that serve as a source of new individuals to populate empty habitat elsewhere, such as patches recovering from fire (box 8.1). But allocating a small percentage of land on the periphery of cities for species preservation, and allotting the rest to human uses, does both humans and other species a disservice. There are a host of practical and ethical reasons why we might choose to support biodiversity within our cities. In fact, even in our most populous cities, a wealth of biodiversity is already present. Wallis Annenberg, president of the Annenberg Foundation, remarked: "It's easy to think of Los Angeles as a concrete jungle. The truth is, we're home to one of the most richly diverse ecosystems in the entire world."[2]

Humans have lived in cities for at least the past five thousand to six thousand years, and some cities such as Damascus have been continuously inhabited for millennia. For much of this time, the sprawl and resource consumption of cities was probably insufficient to directly threaten many species, although urban residents routinely suffered the consequences of deforestation and defaunation of their regional ecosystems. Indeed, for most of our urban history, the size of cities was limited by the natural resources available in the surrounding landscape. Over time, species such as the house mouse (*Mus musculus*), black rat (*Rattus rattus*), and bedbug (*Cimex lectularius*) proliferated in our cities, evolving a dependency on our urban lifestyles.[3]

But now our cities have grown into sprawling metropolitan regions with millions of humans, connected to far-flung places by modern transportation and a global economy. Steel and concrete structures alter habitats, hydrology, temperatures, soils, and nutrients. Many of these alterations come at the expense of native species, particularly in the urban core. Burgeoning megalopolises have unlimited appetites and reach, producing local to planetary consequences for our environment and societies. These consequences manifest when city managers and urban planners treat nature preservation in the city as a luxury that can be ignored or removed when inconvenient. These urban natural amenities, called green infrastructure, are often the first planning elements to be discarded when there are cost overruns or construction difficulties. But this antiquated view of the role of nature and green space in cities has become maladaptive. As Richard Weller describes it: "The city is now *everywhere,* and the world is a hybridized, denatured, co-evolving ecology of our own making. The global city, spread across vast landscapes of resource extraction and waste, is the new nature, and this new nature is suicidal unless we transition cities from their basis in nineteenth-century engineering and move them toward twenty-first-century understandings of ecology."[4]

Green infrastructure is a collection of vegetated spaces, such as parks, wetlands, and street trees. When appropriately designed and connected, green infrastructure can function as a networked system providing critical services, similar to our gray infrastructure of roads, sidewalks, and sewers. Far from optional, green infrastructure is a life support system that makes cities habitable and sustainable. Just as the impacts of a city's gray infrastructure radiate out to the entire region, so do the effects of green infrastructure. Smart urban planning with integrated green infrastructure extends from the city out to the region, integrating gray (streets and sewers), green (green space and habitat), and blue (lakes and rivers) infrastructure systems. When designed well, green infrastructure can provide a plethora of benefits to all species, including ourselves.[5]

Allowing Urban Biodiversity to Thrive

Some species avoid our cities as best they can. Their abundance typically declines with increased human density, and these species become imperiled when their entire distributional range is consumed by cities. We notice these losses and conclude that cities are a lost cause for biodiversity. We therefore focus our conservation efforts in the dwindling wilderness far from the glare of city lights, even as the bright lights of our big cities grow closer to protected areas. Globally, land occupied by cities is expected to triple between 2000 and 2030. Nowhere is this dilemma more pronounced than in Southern California, the most urbanized biodiversity hotspot in the world.[6]

Urbanization need not inevitably result in species extinction. On the contrary, cities host a large number of native species, either concentrated in such habitat remnants as old city parks or integrated throughout the green spaces of a city and its suburbs. For example, Australian cities represent the only habitat left for thirty-nine imperiled species, clinging to existence in schoolyards, airports, roadsides, and golf courses. In a 2014 study of 147 cities across the world, native species comprised more than half of the bird fauna in 54 cities and more than half of plants in 110 cities. Twenty percent of all bird species worldwide can be found in cities—more abundant urban green space translates into more abundant birds. The increased abundance of birds may be due to the green space providing a much-needed sanctuary from pesticide-burdened agricultural areas that often surround cities. Thus, we need to include cities in biodiversity conservation plans and capitalize on the many conservation opportunities that cities offer.[7]

The good news is that we have long known how to integrate habitat into our cities. Urban planners and landscape architects can design green space to be adaptive to the environmental constraints of a site and its intended uses, while ecologists can ensure that green infrastructure design considers biodiversity-dictating relationships: habitat size and amount, habitat connectivity, and the urban matrix.[8]

Our first consideration for urban biodiversity planning is habitat size and amount. Understanding how species respond to the availability of remaining habitat is important to their conservation. Some species require large, intact blocks of habitat for their breeding territories, avoiding habitat edges where predators and unfavorable conditions are dominant. Other species may instead benefit from many small habitat patches, particularly those species that have evolved in a patchy environment. For these species, habitat fragmentation may matter less than overall habitat loss.[9]

Small habitat patches can preserve a significant portion of native biodiversity, sometimes providing habitat that exists nowhere else. In practice, this means that managing many small patches of habitat can be a successful conservation strategy. Conversely, it also means that a single small patch cannot be dismissed as worthless from a conservation standpoint. Small habitat patches do run a greater risk of degradation from feral cats and dogs, invasive species, pollution, noise, and light from their surroundings—what ecologists call edge effects—but these small patches can collectively support a large number of native species, provide connectivity in the landscape, and anchor restoration efforts. The conservation value that each small patch offers depends on such species characteristics as dispersal ability, evolutionary history, and adaptive capacity.[10]

The devaluation of small habitat fragments—as per the NCCP Scientific Review Panel's guidance—incurred a cost to California gnatcatcher conservation efforts back in 1996, when the USFWS granted incidental take permits for gnatcatchers in small coastal sage scrub fragments thought to be of low value. These patches turned out to support "unexpectedly higher densities of gnatcatchers" than the USFWS assumed when it permitted their destruction. As a result, 4 percent of known gnatcatcher pairs were lost when just 1 percent of coastal sage scrub habitat was destroyed. Indeed, small coastal sage scrub patches of 1–4 hectares are routinely used by gnatcatchers for feeding and dispersal and should be included in NCCP reserves.[11]

The effectiveness of habitat preserve networks is also influenced by the network's connectivity. Connectedness depends on the location, quality, and abundance of the habitat patches and on the dispersal behavior of the organisms that are meant to use them. Highly mobile organisms, such as plants with wind-dispersed seeds, birds, flying insects, and medium-sized mammals, such as raccoons and coyotes, generally have an easier time moving among fewer, more isolated habitat patches than organisms that are less mobile. Green spaces scattered throughout a city connected by linear features, such as roadside plantings and riparian strips along streams, can allow species with poor dispersal ability to find larger habitat areas and navigate the preserve network.[12]

Wildlife bridges and underpasses are structural elements that allow animals to navigate road networks safely. In Southern California, these installations have become an important mitigation strategy for new and existing roads and highways. For example, bridges and underpasses increase the safety of mountain lions (*Puma concolor*) on the move. The region's density of multilane, high-speed roadways cause almost 50 percent of mountain lion deaths each year. A planned overpass over U.S. Highway 101 would allow the mountain lions in the Santa Monica Mountains to reach the mountain lions in the Simi Hills; if completed, it would be the largest wildlife overpass in the world. To the southeast, concerns for the small, isolated population of mountain lions in the Santa Ana Mountains stalled a development project that might have jeopardized the utility of proposed overpasses across the I-15 freeway—mountain lions won't use the existing underpasses because of the constant presence of homeless people living in them. Connecting these populations would provide a much-needed boost of genetic diversity for the mountain lions and avoid the severe inbreeding that imperiled the Florida panther (*P. concolor coryi*), a federally endangered subspecies of mountain lion that is also threatened by vehicle collisions. These overpasses and underpasses would also work well for smaller organisms that have an even worse chance of successfully crossing Southern California's highways.[13]

As rapid climate change alters temperature and precipitation gradients, habitat connectivity will be imperative for species seeking out more amenable microclimates. Connectivity is important for the continued presence of urban species, to ensure that they maintain adequate genetic variability at the population level to adapt to changing conditions. For coastal sage scrub species, good habitat connectivity will help them recolonize restored or postfire coastal sage scrub patches and adjust to climate change by moving to cooler, wetter areas northward, toward the coast, or at higher elevations.[14]

So far, I have talked only about the habitat itself, but what surrounds the habitat is equally important. Biodiversity in habitat patches and green space is highly influenced by the mixture of land uses that surround them: the landscape matrix. The landscape matrix denotes all the land that is not intentionally dedicated to supporting biodiversity—residential and industrial areas, highway shoulders, golf courses, and ball fields—but can be supportive if landscaping with native plants is prioritized. For example, urban parks are critical pit-stops for migratory birds on their way to and from their breeding territories. An unfriendly urban matrix around city parks affects the biodiversity inside the parks as much as the recreational use inside the park itself, diminishing the park's utility for species conservation.[15]

Early on, the landscape matrix was ignored by conservation efforts, which prioritized the creation of networks of core preserves and habitat patches. The NCCP Scientific Review Panel mentioned the matrix only once in their conservation guidelines for the program. But softening the matrix between these protected habitat areas, by boosting the use of native plants or reducing such mortality factors as pesticide use and outdoor cats, can make a landscape more hospitable for species moving through and living in it. Because the California gnatcatcher can use several habitat types for feeding and dispersing, softening the urban matrix for the gnatcatcher might be as straightforward as planting more trees and revegetating the banks of streams, rivers, and wetlands.[16]

Research on green infrastructure in cities confirms the conservation benefits of matrix-enhancing efforts, no matter how small. In residential neighborhoods in Chicago, clusters of neighbors who use native plants and fruit-bearing trees support a higher diversity of native birds in their neighborhood when compared to neighborhoods with fewer enthusiastic gardeners. In the aggregate, green-thumbed property owners who add features such as ponds, nest boxes, and bird baths can provide a significant amount of habitat for native species. Green space in any space, including green roofs, vegetated walls and window boxes, and vacant lots, provides feeding, breeding, and dispersal habitat for a variety of bird, mammal, and insect species. Greening abandoned elevated train lines, such as New York City's High Line park and the 606 trail in Chicago, provides attractive recreational green space to neighborhoods and greened corridors for urban species. Biodiversity conservation in the matrix is an important option when governments find it infeasible to strictly protect habitat in the periphery and increase development density in the urban core. Softening the urban matrix provides planners and conservationists some redundancy for restoration failures and new threats. Still, some urban-avoider species will require large areas of natural habitat situated far from human impacts.[17]

Regional planning is as important as citywide planning. Efforts to conserve biodiversity in neighborhoods and cities will require that we pay attention to patterns of biodiversity and habitats at larger scales. For example, during my ten-month sabbatical I recorded every animal I saw in my backyard in Claremont—an intensively landscaped 25 square meters with one large California pepper tree (*Schinus molle*), which, despite the name, is not native to California. I saw one species of mammal, thirty bird species (including a band-tailed pigeon, *Pata-*

gioenas fasciata, a new species for me), two reptile species, six species of butterflies, a bunch of other insects and spiders, and an infinite number of ants. Claremont's abundance of mature street trees is responsible for some of this diversity, but so is the city's proximity to the foothills of the San Gabriel Mountains with its grassland, sage scrub, and chaparral forest, largely protected within the Angeles and San Bernardino National Forests. Therefore, the diversity of my backyard was determined not only by its landscaping but by the availability and diversity of habitats at multiple spatial scales.[18]

Finally, let me address outdoor cats. Keeping pet cats indoors provides safety for small animals as they navigate through urban habitats and boosts the quality of small habitat patches by reducing predation. Cats are among the top sources of wildlife mortality in urban areas worldwide, killing wildlife indiscriminately and increasing disease and parasite transmission. In the United States, outdoor cats kill between 1.3 and 4 billion birds annually and potentially up to 22.3 billion mammals. In Australia, cats kill over 800 million mammals annually. They are notoriously ruthless in dispatching small animals; fewer than a quarter of the animals taken to wildlife rehabilitation centers in the United States and the United Kingdom survived their injuries. A cat's well-being is not improved by outdoor life: keeping cats indoors will also keep your cats safer, since cats get hit by cars, contract diseases, and are killed by bigger predators, too. My two indoor cats personally approved this paragraph.[19]

Nature Boosts Resilience—and Rents

Humans have become increasingly urbanized over the past ten thousand years, during a period of unusually stable climate. Unfortunately, we are entering a time of increased climate variability and our cities need to be adaptive; infrastructure and development designs that worked in our previously stable climate are likely to fail now. Managing cities for resilience requires that we start managing them as systems with social, economic, and ecological subsystems that are dynamic, interdependent, and connected at multiple scales. Globally, urban planners are integrating green infrastructure into new city developments or retrofitting older sections to mitigate flooding, clean the air, filter surface water, and provide shade. For example, Green Alley programs in Chicago, Los Angeles, and other U.S. cities use permeable pavement and vegetation to increase water permeability and control flooding. These installations also increase residents' access to green space and boost alley safety and walkability. Green space benefits broader scales, too. In New England, the addition of both public and private protected areas boosted employment numbers, chiefly through increased recreation, without negatively affecting other aspects of local economies. Smarter planning boosts urban resilience to disturbances and environmental change that originates within the city and from larger scales.[20]

One barrier to adding public green space to cities is the mistaken belief that these areas reduce the municipal tax base by restricting development and cost more to establish and maintain than justified by the benefits. The opposite is true. Green space often boosts property values around it and encourages infill into abandoned areas. Indeed, the cost of development that destroys green space is often underestimated, and the new tax revenues gained from poorly planned development may not offset the cost of lost green space and ecosystem services. In Southern California, for example, the loss of half of the region's urban trees to invasive pests represented a ten-year, $616.6 million loss in ecosystem services provided by those trees. Street-level and rooftop green space supports insects, such as bees and wasps, that provide pollination services that are difficult and costly for us to replicate. Of course, more development also risks degrading existing green space and increasing its isolation from other open areas; therefore, long-term plans for green space connectivity and maintenance are essential.[21]

Green infrastructure projects frequently increase the housing prices around them and, thus, affect the affordability of neighborhoods. The paradox of creating green space for lower-income residents, which ultimately drives gentrification and displaces them, is a phenomenon observed worldwide. Without equity policies, residents who cannot afford the increased rents are forced to move to areas with fewer natural amenities, institutionalizing green space inequities in a city. Gentrification driven by green space can appear to be a planned outcome, meant to increase real estate values and tax revenues in neighborhoods with desirable locations—in the case of New York City's Central Park, this was the expressed intent. The city's High Line project boosted nearby property values by over 100 percent and encouraged $2 billion in new development.[22]

The risk of environmental gentrification increases when cities engage in public-private partnerships to green up existing neighborhoods, the consequence of governments delegating their responsibility for the provision of public goods. Environmental nonprofit organizations often serve as the private partner in these arrangements, and they may not consider social impacts as part of their mission. In the case of the 606 trail in Chicago, the initial requests for the project came from the neighborhood's lower-income residents, who recognized their lack of green space and recreational areas. These residents were ultimately displaced by the higher real estate values generated by the completed project. To avoid this outcome, planners could use smaller, more numerous parks to add green space to neighborhoods without driving up housing costs. These pocket parks can also increase green space connectivity for less mobile species and soften the urban matrix. Ultimately, to avoid damaging residents' trust in gov-

ernment through unsustainable gentrification projects, green infrastructure projects require purposeful decisions, involving all stakeholders, regarding the types and amounts of green space needed and
how these green spaces will contribute to a diversity of social and
ecological needs.[23]

Vitamin G(reen) and Caring for Nature

Humans didn't evolve in a concrete jungle—we evolved in savan-
nas, woodlands, forests, and wetlands. It stands to reason, therefore,
that our physical and mental well-being is improved by significant
amounts of green space. This realization is not new. In 1865, Frederick Law Olmsted, the famous nineteenth-century designer of urban
green space such as Central Park in New York City, remarked: "It is a
scientific fact that the occasional contemplation of natural scenes of
an impressive character ... is favorable to the health and vigor of men.
[The government has a duty to assure that] enjoyment of the choicest
natural scenes in the country and the means of recreation connected
with them [be] laid open to the use of the body of the people." Frederick Law Olmsted Jr., a landscape architect like his father, felt similarly about the importance of nature and helped write the legislation
that initiated the National Park Service. Indeed, protected areas are
prime locations for recreation and tourism today, precisely for their
value to mental and physical well-being. But the benefits of green
spaces aren't restricted to leisurely recreation; participating in trash
pick-ups and invasive plant species pulls in parks provides both educational and recreational opportunities.[24]

Green space benefits physical health, such as through reduced air
and water pollution, but green space also improves our mental health.
In 2006, Peter Groenewegen and his colleagues coined the term *Vitamin G,* with a "G" for green space, highlighting the importance of

nature to the well-being of urban residents. Health and well-being are improved through stress reduction, increased physical activity, and improved feelings of social connectedness. In Toronto, Canada, ten extra trees per block was correlated with a marked decrease in heart disease and made people feel $10,000 richer and seven years younger. In Japan, the act of walking through a forest for health effects is called *shinrin-yoku,* which translates to "forest bathing." These forest baths have been shown to reduce stress hormones and blood pressure. Even one forest bath a month can provide a measurable boost to the immune system. A Dutch team surveyed more than ten thousand people in the Netherlands about their health and the green space around them: people in greener neighborhoods reported better physical and mental health. In China, higher neighborhood green space was correlated with lower mortality rates for elderly residents over a fourteen-year span. Even simply increasing tree canopy cover can reduce neighborhood-level crime rates. In Cincinnati, where an invasive beetle called the emerald ash borer (*Agrilus planipennis*) has been killing ash trees since 2007, crime rates increased in neighborhoods where tree cover had declined from ash die-off. Around the world, neighborhoods with more green space experience fewer crimes, less aggression, and less mental fatigue and stress.[25]

Green space benefits children, too. Children in industrialized nations are experiencing a condition that Richard Louv describes as "nature-deficit disorder," with a range of symptoms linked to a decline in access to green space—a consequence of safety concerns, lack of nearby green space, and the allure of video games and other electronics. Children without routine access to nature can experience a cluster of physical ailments, learning disabilities, and cognitive development delays during childhood, as well as lowered support for environmental protection and engagement with nature as adults. In Denmark, children growing up in areas devoid of green space face up to twice the risk of developing a mental illness than children who grow up in neighborhoods with high amounts of green space. Routine exposure and proximity to nature may reduce acute and chronic stress through multiple mechanisms, including physiological improvements with access to areas to play and exercise and psychological improvements from social gathering areas with decreased noise.[26]

Some cities have heeded these results. As one of the pilot cities in the Cities Connecting Children to Nature program administered by the National League of Cities, Austin, Texas, recently adopted a Children's Outdoor Bill of Rights. The list includes a child's right to climb trees, catch fish, and see the stars. Austin and the other seventeen participating cities can access millions of dollars in funding for urban green space projects in their cities.[27]

In 1984, Edward O. Wilson published a transformative book, *Biophilia,* detailing his hypothesis that human exposure to nature increases our desire to protect it. Conversely, humans with little exposure to the natural world are less concerned with saving it or perhaps even grow to fear it. His hypothesis originated from personal experience; he attributed his lifelong study of insects and their behaviors to his childhood days spent outdoors in Alabama. Urban planning expert Timothy Beatley later extended Wilson's hypothesis to "biophilic cities," an urban design approach that emphasizes the ability of a city resident to experience nature within walking distance of their home. I can relate to this biophilia concept. The time I spent with my childhood friends in Chicago's Cook County Forest Preserves — riding bikes, looking for turtles, and building tree forts and fake beaver dams — prompted my career as an ecologist. Exposure to nature influences how well we look after the environment and each other and how much we are willing to sacrifice for the welfare of future generations.[28]

Reconciling Our Cities with Nature

In his 2003 book *Win-Win Ecology,* noted ecologist Michael Rosenzweig explains that conservation consists of three broad approaches; conveniently, they all start with an "R." All contribute to biodiversity conservation in urbanizing regions.[29]

The first is *reservation* ecology—setting aside ecologically valuable areas in parks and preserves. Now-familiar HCPs and NCCP plans engage in reservation ecology. Fencing off and protecting biodiversity in parks and preserves has been a standard practice for more than a century (see chapter 5), but this approach has significant problems and limitations. First, some parks are established by removing indigenous communities from their territories, thus creating a conflict between biodiversity conservation and the rights of indigenous cultures with extensive ecological knowledge of the place. For example, Native Americans were forcibly removed from many of the early national parks in the United States, and indigenous communities are routinely removed from protected areas around the world today. Second, climate change and invasive species are expected to have a significant impact on protected areas, potentially limiting their ability to preserve biodiversity in the long term. Finally, protected areas are rarely large enough to encompass entire ecosystems or wide-ranging species. Even though Everglades National Park encompasses over 6,000 square kilometers, the massive park only protects the bottom third of the Everglades ecosystem. The famous seasonal migrations of African wildlife pass through multiple countries, far beyond the boundaries of existing national parks.[30]

The second R is *restoration* ecology—degraded or destroyed areas are triaged and replanted, and then they are protected to allow native habitats to recover. I haven't spoken much about habitat restoration in this book, but restoration efforts are included in some NCCP plans, and restoring habitat for federally listed species is a routine part of many recovery plans. Restoration also happens in former industrial areas called brownfields, vacant lots, and other abandoned areas where city planners and engaged locals seek to return nature and beauty to a neighborhood. Just as with protected areas, habitats restored improperly or surrounded by a hostile matrix risk becoming ecological traps for some species, attracting individuals that suffer higher mortality and lower reproductive success. Although we have limited experience with coastal sage scrub restoration, existing studies suggest that it is possible to restore small patches to a quality that is useful to California gnatcatchers for foraging and dispersal, with mixed success for other species.[31]

The third approach—*reconciliation*—is completely absent from the HCP and NCCP conservation approaches for the California gnatcatcher. Also called biodiversity-sensitive urban design or conservation development, reconciliation in cities entails softening the urban matrix to permit other species to coexist along with us. New York City's High Line park is a reconciliation project, incorporating ecological succession and wildness into the park's design. Reconciliation must involve local communities, since their perception of green spaces heavily influences how successfully these spaces will be maintained and enjoyed. Without an education and outreach component, neighbors may respond negatively to green space projects that look "messy," intimating neighborhood neglect and safety issues. Working collaboratively, ecologists, landscape architects, and planners can guide restored lots to support biodiversity in ways that are more appealing to the local community.[32]

Coalitions of ecologists, urban planners, and landscape architects need to collectively influence urban planning and reconcile humans with their natural environments. Local stakeholders must be involved to assure that new green space does not elevate real estate prices beyond what locals can afford. Green infrastructure plans cannot focus solely on land within the city limits but must also consider the region. Regional reconciliation plans need to protect remaining wilderness and intact ecosystems on the urban periphery—these are the reservoirs for genetic and population diversity that support biodiversity and ecosystem functioning within the city. Urban and regional planners must either design gray and green infrastructures concurrently or start with the green infrastructure system and plan development and the gray infrastructure network around it.[33]

Reconciliation in Southern California

Most of this book discusses reservation-based conservation in Southern California by way of the NCCP process. Restoration of coastal sage scrub, river riparian zones, and other habitats is an ongoing thrust across the region, too; these will dramatically increase the connectivity of the habitat patches protected by the NCCP policy. And evidence is growing of our successful reconciliation with urban nature in Southern California and elsewhere.

After decades of absence, the federally threatened western snowy plover—one of the species threatened by the El Morro Village mobile home park—is nesting again on Southern California beaches, thanks to local efforts by cities and environmental groups that fenced off nesting areas and restored the beach vegetation. The vegetation and debris left on the restored beaches may also help combat beach erosion due to sea level rise and high-intensity storms. Since its ESA listing in 1993, the plover's recovery has already achieved its population target for recovery, decades before it was predicted to do so. Farther east, a pair of federally endangered piping plovers (*Charadrius melodus*) successfully reared chicks on one of Chicago's busiest beaches after a sixty-year absence, also after extensive efforts to fence off and protect the pair's nesting area.[34]

Despite the constant din of jet engines, several protected species have returned to the Los Angeles International Airport (LAX) Dunes Preserve, a fenced 122-hectare patch of land sandwiched between the Pacific Ocean and the western end of two LAX runways. Burrowing owls and California gnatcatchers are now breeding there, as are El Segundo blue butterflies. Managers hope to reintroduce the critically endangered Pacific pocket mouse to the preserve soon.[35]

Sierra Nevada bighorn sheep (*Ovis canadensis sierrae*) are making a comeback in Riverside and San Diego Counties after their endangered listing under the ESA in 1996, when their population had dropped from over 1,000 sheep to just 271. Now more than 900 individuals, the subspecies is numerous enough to generate lawsuits between conservationists who want golf courses fenced (to prevent sheep from drowning in water features) and golf club owners who don't mind seeing them munching along the fairways.[36]

After over a century of persecution, urban coyotes have reclaimed their territory and are integrating themselves back into urban ecosystems. Regaining their rightful place in Southern California communities is likely to help numerous species and habitat restoration efforts. A now-classic study by Kevin Crooks and Michael Soulé demonstrated that birds in small coastal sage scrub patches suffered higher mortality from feral cats than birds in larger patches; coyotes (noted cat-killers) were more common in the larger patches and kept cat numbers low.[37]

In Ventura County, the board of supervisors recently approved a county ordinance with multiple provisions to maintain habitat connectivity for wildlife. The provisions include a 61-meter no-development buffer zone along streams and rivers, reduced outdoor lighting next to habitat corridors, and development restrictions in the Tierra Rejada Valley, which connects the Santa Monica Mountains and Los Padres National Forest. Despite its exclusion from the NCCP pilot plans, Ventura County has forged ahead with its own open-space and habitat-protection policies, keeping pace with the region.[38]

Alas, not all reconciliation efforts are going smoothly. In Los Angeles County, development fees established under Measure A—called Quimby funds because they are used at the mayor's discretion—have funded a surge of park development with mixed results. Starting in 2012, the 50 Parks Initiative began to develop small pocket parks throughout the downtown area and nearby neighborhoods. The goal of the initiative is to develop fifty-one parks for a total of 73 hectares, which will remain permanently protected in the city charter. But the addition of these many small parks will be offset by the loss of a large section of Exposition Park for filmmaker George Lucas's new museum of narrative art.[39]

During my sabbatical in Southern California, I tried to visit as many naturally vegetated open space areas as I could. As my field notes attest (see boxes 1.1–8.1), nearly every park I visited was crowded with people, regardless of the time of day or day of the week. Obviously, these parks and open spaces are in high demand. If properly managed, the parks can satisfy a variety of recreational needs and ecosystem services for humans while contributing to biodiversity conservation. Thanks to laws such as the 1991 NCCP Act and the 1976 California Coastal Act—which protects and maintains public access to beaches—the beauty and diversity of California's landscapes can be enjoyed by residents and tourists alike. While the NCCP policy itself hasn't spread to other states, multispecies, habitat-level conservation approaches are becoming far more common, whether under the ESA (MSHCPs) or through market mechanisms such as habitat conservation banking. Other than the NCCP policy's surrogate species, there is nothing so distinctive about the approach that it cannot be adapted to other regions, habitats, and landscapes.

In another thirty years, we may determine that the California gnatcatcher was not an optimal or even adequate surrogate species for habitat protection and restoration in Southern California, particularly in the face of climate change, invasive species, and relentless development. But the gnatcatcher does have the ability to capture the attention and sympathy of the public, and caring about nature—or even just whiny gray fluff balls—is key to reconciliation efforts. Reconciliation begins when humans care enough about nature to save it. Caring manifests itself in the form of funding, habitat set-asides, well-planned green infrastructure, and effective policy, of course. But these are possible only when residents demand nature-sparing regulations and urban planning from their local governments.[40]

Epilogue

When my sabbatical began, I requested email alerts on the California gnatcatcher from the *Federal Register,* the daily report of the U.S. executive branch and an important source of information for this book. Federal agencies such as the USFWS are required to publish actions, decisions, and solicitations for public comments in the *Federal Register.* As I was finishing the draft of this book, I received a *Federal Register* notice on June 19, 2019: "Habitat Conservation Plan for the Coastal California Gnatcatcher; Categorical Exclusion for 93-129 Ltd, Orange County, California." The USFWS was soliciting public comments on an incidental take permit for a development project—a project that nicely encapsulated the questions that remain unanswered by this book.[1]

A 20-hectare development project on parcel #93-129 in Laguna Niguel sought a ten-year incidental take permit for California gnatcatchers—up to two known pairs and 1.7 hectares of their habitat. The take would be mitigated by setting aside 5.2 hectares of coastal sage scrub, 4.3 hectares of which didn't yet exist, but would be created and restored at some later date elsewhere. The project included only four houses.

Four houses on 20 hectares in the city of Laguna Niguel? The city is a carpet of subdivisions—arcing streets lined with tightly packed houses on 0.1-hectare lots. Thin ribbons of green are relegated to the ridgetops and canyons. Where could such a low-density housing project possibly fit in such a densely urbanized city? A search of the parcel number yielded Bear Brand Ranch, a gated community of luxury homes perched along a ridgetop with views of the Pacific Ocean, the Santa Ana Mountains, and the frothy urban masses below. I had a hard time believing that the addition of four multi-million-dollar estates was a wise decision during a severe affordable housing crisis.[2]

The notice indicated that this was the second ten-year incidental take permit for the California gnatcatcher that the developer had sought. The first permit was issued by the USFWS on June 25, 2007, right before the 2008 financial crisis that brought the Southern California development industry to its knees. The economic slump plus some bureaucratic issues — none of which involved the gnatcatcher or other listed species — delayed the project past the expiration date of the first incidental take permit, and now the developer was asking for another ten years.

A great deal of wrath and worry has been laid at the feet of protected species such as the California gnatcatcher and the Stephens' kangaroo rat. Conservation critics claim that their protection eliminates jobs, delays much-needed new housing, and risks damage to the regional economy. But this development project was delayed for ten years by a national economic crisis and jurisdictional issues that the California gnatcatcher had no part in. The first incidental take permit was granted, as was the second permit. California gnatcatchers are not idling any bulldozers here.[3]

It is possible that the gnatcatchers displaced by this development will survive the destruction of their habitat and thread their way through the remaining slim fingers of green space to find a new home. But I can't help asking: four houses gained and four gnatcatchers lost — is that a smart trade? As a consequence of the project's off-site mitigation plan, any remaining habitat in this neighborhood will be smaller in size and even more isolated from the region's coastal sage scrub, and nearby residents will lose their proximity to a natural habitat supporting a federally listed species. The NCCP policy was based on the belief that endangered species, such as the California gnatcatcher, were creating development bottlenecks, housing shortages, and economic hardship. But this permit renewal suggests that larger forces may be equally to blame. We can coexist with native species in our urban landscapes, and we will benefit in many ways by doing so, but only if we are honest about the risks and benefits that these species represent.

Notes

Abbreviations

BC	*Biological Conservation*
CB	*Conservation Biology*
ELR	*Environmental Law Reporter*
LAT	*Los Angeles Times*
PNAS	*Proceedings of the National Academy of Sciences of the United States of America*
PRSB	*Proceedings of the Royal Society B*
SELS	Stanford Environmental Law Society
WB	*Western Birds*

Chapter 1. Setting the Scene

1. Peter U. Clark et al., "Global Climate Evolution during the Last Deglaciation," *PNAS* 109 (2012): E1134–E1142; Julia Rosen, "A Wrinkle in Time: Finding the Ice Age in Urban Los Angeles," *LAT,* June 21, 2014; Associated Press, "Los Angeles Subway Work Uncovers Array of Ice Age Fossils," *NBC News Local 4,* Dec. 11, 2017, https://www.nbclosangeles.com/news/local/Los-Angeles-Subway-Work-Uncovers -Array-of-Ice-Age-Fossils-463492993.html; George Monbiot, *Feral: Rewilding the Land, the Sea, and Human Life* (Chicago: University of Chicago Press, 2014); A. M. Lister and A. V. Sher, "Evolution and Dispersal of Mammoths across the Northern Hemisphere," *Science* 350 (2015): 805–809; Lindsey T. Yann et al., "Dietary Ecology of Pleistocene Camelids: Influences of Climate, Environment, and Sympatric Taxa," *Palaeogeography Palaeoclimatology Paleoecology* 461 (2016): 389–400; La Brea Tar Pits and Museum website: https://tarpits.org.

2. Joe Roman, *Listed: Dispatches from America's Endangered Species Act* (Cambridge, Mass.: Harvard University Press, 2011), 36–37; Larisa R. G. DeSantis et al., "Implications of Diet for the Extinction of Saber-Toothed Cats and American Lions," *PLoS ONE* 7 (2012): https://doi.org/10.1371/journal.pone.0052453; Monbiot, *Feral;*

B. Figueirido, P. Palmqvist, and J. A. Pérez-Claros, "Ecomorphological Correlates of Craniodental Variation in Bears and Paleobiological Implications for Extinct Taxa: An Approach Based on Geometric Morphometrics," *Journal of Zoology* 277 (2009): 70–80; Shelly L. Donohue et al., "Was the Giant Short-Faced Bear a Hyper-Scavenger? A New Approach to the Dietary Study of Ursids Using Dental Microwear Textures," *PLoS ONE* 8 (2013): https://doi.org/10.1371/journal.pone.0077531; William A. Akersten, Christopher A. Shaw, and George T. Jefferson, "Rancho La Brea: Status and Future," *Paleobiology* 9 (1983): 211–217; Elizabeth A. Hadly and Robert S. Feranec, "Paleovertebrate Communities," in *Ecosystems of California,* ed. Harold Mooney and Erika Zavaleta (Oakland: University of California Press, 2016), 155–168.

3. Clark et al., "Global Climate Evolution"; Jeremy D. Shakun et al., "Global Warming Preceded by Increasing Carbon Dioxide Concentrations during the Last Deglaciation," *Nature* 484 (2012): 49–54; Akersten, Shaw, and Jefferson, "Rancho La Brea"; M. Kat Anderson, Michael G. Barbour, and Valerie Whitworth, "A World of Balance and Plenty: Land, Plants, Animals, and Humans in a Pre-European California," *California History* 76 (1997): 12–47.

4. Jacquelyn L. Gill et al., "Pleistocene Megafaunal Collapse, Novel Plant Communities, and Enhanced Fire Regimes in North America," *Science* 326 (2009): 1100–1103; Charles C. Mann and Mark L. Plummer, *Noah's Choice: The Future of Endangered Species* (New York: Alfred A. Knopf, 1995), 46–47; Roman, *Listed,* 30–31; Christopher Sandom et al., "Global Late Quaternary Megafauna Extinctions Linked to Humans, Not Climate Change," *PRSB* 281 (2014): https://doi.org/10.1098/rspb.2013.3254; David J. Meltzer, "Pleistocene Overkill and North American Mammalian Extinctions," *Annual Review of Anthropology* 44 (2015): 33–53; David Tilman et al., "Future Threats to Biodiversity and Pathways to Their Prevention," *Nature* 546 (2017): 73–81; Connie J. Mulligan, Andrew Kitchen, and Michael M. Miyamoto, "Updated Three-Stage Model for the Peopling of the Americas," *PLoS ONE* 3 (2008): https://doi.org/10.1371/journal.pone.0003199; Todd J. Braje et al., "Finding the First Americans," *Science* 358 (2017): 592–594; C. L. Scheib et al., "Ancient Human Parallel Lineages within North America Contributed to a Coastal Expansion," *Science* 360 (2017): 1024–1027; Traci Watson, "Is Theory about Peopling of the Americas a Bridge Too Far?" *PNAS* 114 (2017): 5554–5557; Lizzie Wade, "Ancient DNA Tracks Migrations around Americas," *Science* 362 (2018): 627–628; L. Mark Raab, "Political Ecology of Prehistoric Los Angeles," in *Land of Sunshine: An Environmental History of Metropolitan Los Angeles,* ed. William Deverell and Greg Hise (Pittsburgh: University of Pittsburgh Press, 2005), 23–37; Terry L. Jones and Kacey Hadick, "Indigenous California," in Mooney and Zavaleta, *Ecosystems of California,* 173.

5. Akersten, Shaw, and Jefferson, "Rancho La Brea"; Rosen, "Wrinkle in Time"; DeSantis et al., "Implications of Diet"; Monbiot, *Feral;* Tilman et al., "Future Threats to Biodiversity"; Watson, "Theory about Peopling"; T. L. Jones et al., "The Protracted Holocene Extinction of California's Flightless Sea Duck (*Chendytes lawi*) and Its Implications for the Pleistocene Overkill Hypothesis," *PNAS* 105 (2008): 4105–4108; Jones and Hadick, "Indigenous California," 181. "Chief architects" refers to the overkill hypothesis in which overhunting by newly arrived humans into Australia and the Western Hemisphere drove megafauna species extinct. Jones and Hadick, "Indigenous California," 178, maintain that there is little evidence to support this hypothesis in Southern California.

6. Robert M. Zink and Rachelle C. Blackwell, "Molecular Systematics and Biogeography of Aridland Gnatcatchers (Genus *Polioptila*) and Evidence Supporting Species Status of the California Gnatcatcher (*Polioptila californica*)," *Molecular Phylogenetics and Evolution* 9 (1998): 26–32; Brian Tilston Smith et al., "Species Delimitation and Biogeography of the Gnatcatchers and Gnatwrens (Aves: Polioptilidae)," *Molecular Phylogenetics and Evolution* 126 (2018): 45–57; La Brea Tar Pits, "Our Collections: Flora," https://tarpits.org/research-collections/tar-pits-collections/botany-collections; Anderson, Barbour, and Whitworth, "World of Balance and Plenty."

7. Kristina Hill, "Form Follows Flows: Systems, Design, and the Aesthetic Experience of Change," in *Nature and Cities: The Ecological Imperative in Urban Design and Planning,* ed. Frederick R. Steiner, George F. Thompson, and Armando Carbonell (Cambridge, Mass.: Lincoln Institute of Land Policy, 2016), 345–359; Carey McWilliams, *Southern California: An Island on the Land* (Salt Lake City, Utah: Peregrine Smith Books, 1983); Anderson, Barbour, and Whitworth, "World of Balance and Plenty"; William S. Simmons, "Indian Peoples of California," in *Contested Eden: California before the Gold Rush,* ed. Ramón A. Gutiérrez and Richard J. Orsi (Berkeley: University of California Press, 1998), 48–77; Raab, "Political Ecology"; Paula M. Schiffman, "The Los Angeles Prairie," in Deverell and Hise, *Land of Sunshine,* 38–51; Benjamin Madley, *An American Genocide: The United States and the California Indian Catastrophe, 1846–1873* (New Haven: Yale University Press, 2016), 18; Jones and Hadick, "Indigenous California," fig. 10.1; Schiffman, "Los Angeles Prairie"; Anderson, Barbour, and Whitworth, "World of Balance and Plenty"; Jon E. Keeley, "Native American Impacts on Fire Regimes of the California Coastal Ranges," *Journal of Biogeography* 29 (2002): 303–320; Glenn J. Farris, "Depriving God and the King of the Means of Charity," in *Indigenous Landscapes and Spanish Missions,* ed. Lee Panich and Tsim D. Schneider (Tucson: University of Arizona Press, 2014), 135–153; Anna Klimaszewski-Patterson and Scott A. Mensing, "Multi-Disciplinary Approach

to Identifying Native American Impacts on Late Holocene Forest Dynamics in the Southern Sierra Nevada Range, California, USA," *Anthropocene* 15 (2016): 37–48; William Preston, "Serpent in the Garden: Environmental Change in Colonial California," *California History* 76 (1997): 260–298; Mike Davis, *Ecology of Fear: Los Angeles and the Imagination of Disaster* (New York: Vintage Books, 1998), 21–23; Terry L. Jones et al., "Demographic Crises in Western North America during the Medieval Climatic Anomaly," *Cultural Anthropology* 40 (1999): 137–170.

8. Sara Kiley Watson, "8,200 Years Ago, California's Forecast Was 150 Years of Rain," *Popular Science,* June 28, 2017, https://www.popsci.com/cave-deposits-california-rain/; Frank W. Davis, Peter A. Stine, and David M. Stoms, "Distribution and Conservation Status of Coastal Sage Scrub in Southwestern California," *Journal of Vegetation Science* 5 (1994): 743–756; Anderson, Barbour, and Whitworth, "World of Balance and Plenty"; J. C. Burger et al., "Restoring Arthropod Communities in Coastal Sage Scrub," *CB* 17 (2003): 460–467; Jonathan L. Atwood, "Speciation and Geographic Variation in Black-Tailed Gnatcatchers," *Ornithological Monographs* 42 (1988): 1–74.

9. McWilliams, *Southern California*; Simmons, "Indian Peoples"; Madley, *American Genocide;* Thomas Curwen, "Reviving L.A.'s Native Tongue and Finding a Forgotten World; Tongva Language Gives Students a Link to the Past," *LAT,* May 11, 2019; Preston, "Serpent in the Garden"; Douglas C. Sackman, *Orange Empire: California and the Fruits of Eden* (Berkeley: University of California Press, 2007); Farris, "Depriving God"; Raab, "Political Ecology"; Rosanne Welch, "A Brief History of the Tongva Tribe" (Ph.D. diss., Claremont Graduate University, 2006); James N. Gregory, "Review: Inventing California," *Reviews in American History* 13 (1985): 570–576; David Vogel, *California Greenin': How the Golden State Became an Environmental Leader* (Princeton, N.J.: Princeton University Press, 2018); Iris H. W. Engstrand, "Seekers of the 'Northern Mystery,'" in Gutiérrez and Orsi, *Contested Eden,* 78–110.

10. Welch, "Brief History"; R. F. Heizer and M. A. Whipple, *The California Indians: A Source Book* (Berkeley: University of California Press, 1971), 355; Sackman, *Orange Empire.*

11. Welsh, "Brief History"; McWilliams, *Southern California;* Madley, *American Genocide,* 26–29, 34.

12. McWilliams, *Southern California;* Steven W. Hackel, "Land, Labor, and Production," in Gutiérrez and Orsi, *Contested Eden,* 111–146; Schiffman, "Los Angeles Prairie"; Preston, "Serpent in the Garden"; Anderson, Barbour, and Whitworth, "World of Balance and Plenty"; Hackel, "Land, Labor, and Production"; Vogel, *California Greenin'.*

13. McWilliams, *Southern California;* Hackel, "Land, Labor, and Production"; Welch, "Brief History"; Stephanie S. Pincetl, *Transforming California: A Political History of Land Use and Development* (Baltimore: John Hopkins University Press, 1999).

14. McWilliams, *Southern California;* Karen Clay and Werner Troesken, "Ranchos and the Politics of Land Claims," in Deverell and Hise, *Land of Sunshine,* 52–66; Hackel, "Land, Labor, and Production"; Madley, *American Genocide,* 156–157.

15. Clay and Troesken, "Ranchos"; Sackman, *Orange Empire,* 27; quote in McWilliams, *Southern California,* 206; Martin Wachs, "Autos, Transit, and the Sprawl of Los Angeles—the 1920s," *Journal of the American Planning Association* 50 (1984): 297–310; Jennifer Wolch, Manuel Pastor Jr., and Peter Dreier, "Introduction," in *Up against the Sprawl: Public Policy and the Making of Southern California,* ed. Wolch, Pastor, and Dreier (Minneapolis: University of Minnesota Press, 2004), 4; Blake Gumprecht, "Who Killed the Los Angeles River?" in Deverell and Hise, *Land of Sunshine,* 115–134.

16. Wachs, "Autos, Transit, and Sprawl"; William Fulton, *The Reluctant Metropolis: The Politics of Urban Growth in Los Angeles* (Baltimore: Johns Hopkins University Press, 2001). The term *open space* used here and elsewhere implies land that is not dedicated to agricultural, industrial, or urban land uses. But where I talk about open space as specifically set aside or zoned as such, I refer to a more specific term, which means that it is a publicly owned property for recreation and conservation (called "fee protected" open space), as used in Maria João Santos, Terry Watt, and Stephanie Pincetl, "The Push and Pull of Land Use Policy: Reconstructing 150 Years of Development and Conservation Land Acquisition," *PLoS ONE* 9 (2014): e103489.

17. Wachs, "Autos, Transit, and Sprawl."

18. Wachs, "Autos, Transit, and Sprawl"; Fulton, *Reluctant Metropolis,* 7–8; Wolch, Pastor, and Dreier, "Introduction," 6–7.

19. Daniel Press, *Saving Open Space: The Politics of Local Preservation in California* (Berkeley: University of California Press, 2002), 33–38; Santos, Watt, and Pincetl, "Push and Pull."

20. McWilliams, *Southern California;* George W. James, *California, Romantic and Beautiful* (Boston: Page, 1914), 250, 252; Sackman, *Orange Empire;* Joseph Grinnell, "Birds of the Pacific Slope of Los Angeles County," *Pasadena Academy of Sciences* 11 (1898): 1–52; Joseph Grinnell and Alden H. Miller, "The Distribution of the Birds of California," *Pacific Coast Avifauna,* no. 27 (Berkeley, Calif.: Cooper Ornithological Society, 1944).

21. James, *California,* 264, 359; Davis, *Ecology of Fear,* 76–80; McWilliams, *Southern California;* Wachs, "Autos, Transit, and Sprawl"; Wolch, Pastor, and Dreier,

"Introduction," 7; Robert Gottlieb, *Reinventing Los Angeles: Nature and Community in the Global City* (Cambridge, Mass.: MIT Press, 2007), 173; Stephanie Pincetl, Terry Watt, and Maria J. Santos, "Land Use Regulation for Resource Conservation," in Mooney and Zavaleta, *Ecosystems of California,* 899–924; Gottlieb, *Reinventing Los Angeles,* 29; U.S. Census Bureau, 2019; Sackman, *Orange Empire.* For example, Heather L. Hulton VanTassel et al., "Environmental Change, Shifting Distributions, and Habitat Conservation Plans: A Case Study of the California Gnatcatcher," *Ecology and Evolution* 7 (2017): 10326–10338.

22. Wachs, "Autos, Transit, and Sprawl"; Fulton, *Reluctant Metropolis,* 61; Press, *Saving Open Space,* 38–47; Pincetl, Watt, and Santos, "Land Use Regulation"; Pascale Joassart-Marcelli, William Fulton, and Juliet Musso, "Can Growth Control Escape Fiscal and Economic Pressures? City Policy before and after the 1990s Recession," in Wolch, Pastor, and Dreier, *Up against the Sprawl,* 255–277.

23. Fulton, *Reluctant Metropolis,* 175–176.

24. Pincetl, Watt, and Santos, "Land Use Regulation."

25. Stephanie Pincetl, "The Preservation of Nature at the Urban Fringe," in Wolch, Pastor, and Dreier, *Up against the Sprawl,* 225–251; Pincetl, Watt, and Santos, "Land Use Regulation"; Craig W. Thomas, *Bureaucratic Landscapes: Interagency Cooperation and the Preservation of Biodiversity* (Cambridge, Mass.: MIT Press, 2003), 131; Fulton, *Reluctant Metropolis,* 155, 172.

26. Fulton, *Reluctant Metropolis,* 226; Press, *Saving Open Space,* 6, 11, 42–43; Santos, Watt, and Pincetl, "Push and Pull"; Wolch, Pastor, and Dreier, "Introduction," 8; Pincetl, "Urban Fringe" (for the conservation paradox of Prop 13, see p. 901); Pincetl, Watt, and Santos, "Land Use Regulation."

27. Pincetl, Watt, and Santos, "Land Use Regulation," 916; Wolch, Pastor, and Dreier, "Introduction," 12; Fulton, *Reluctant Metropolis,* 283. Pincetl, "Urban Fringe," suggests that this dynamic can be better understood under an "urban regime theory" conceptualization.

28. Like the California gnatcatcher, indigenous tribes rely on legal recognition for preservation: Alexa Koenig and Jonathan Stein, "Federalism and the State Recognition of Native American Tribes: A Survey of State-Recognition Processes across the United States," *Santa Clara Law Review* 48 (2008): 79–153; Welch, "Brief History"; Curwen, "Reviving L.A.'s Native Tongue." Jennifer Rice Epstein, "Gage Mansion Is Moldering Away Right in Front of Its Owners' Eyes," *Los Angeles Magazine,* July 10, 2017; Mikhail Chester et al., "Parking Infrastructure: A Constraint on or Opportunity for Urban Redevelopment? A Study of Los Angeles County Parking Supply and Growth," *Journal of the American Planning Association* 81 (2015): 268–286.

29. Press, *Saving Open Space*, 16–17, 63.

30. Santos, Watt, and Pincetl, "Push and Pull"; Pincetl, Watt, and Santos, "Land Use Regulation." Regarding the Carlsbad Highlands Reserve: Deborah Sullivan Brennan, "Wildlife Officials, Mountain Bikers Fight Turf War over Carlsbad Reserve," *San Diego Union-Tribune*, Apr. 13, 2019.

31. Daniel M. Evans et al., "Species Recovery in the United States: Increasing the Effectiveness of the Endangered Species Act," *Issues in Ecology* 20 (2016): 1–28; State of California, Natural Resources Agency, "State and Federally Listed Endangered, Threatened, and Rare Plants of California," Jan. 2, 2020, https://nrm.dfg.ca.gov/FileHandler.ashx?DocumentID=109390&inline; State of California, Natural Resources Agency, "State and Federally Listed Endangered and Threatened Animals of California," Aug. 7, 2019, https://nrm.dfg.ca.gov/FileHandler.ashx?DocumentID=109405&inline; Helen M. Regan et al., "Species Prioritization for Monitoring and Management in Regional Multiple Species Conservation Plans," *Diversity and Distributions* 14 (2008): 462–471; Bernie Tershy et al., "Biodiversity," in Mooney and Zavaleta, *Ecosystems of California*, 187–212; Michael Hiltzik, "The Delta Smelt Heads for Extinction, Marking a Half-Century of Failed California Water Policy," *LAT*, Jan. 5, 2018; Jeffrey E. Lovich and Joshua R. Ennen, "Assessing the State of Knowledge of Utility-Scale Wind Energy Development and Operation on Non-Volant Terrestrial and Marine Wildlife," *Applied Energy* 103 (2013): 52–60; Schiffman, "Los Angeles Prairie"; Thomas, *Bureaucratic Landscapes;* Brett Johnson, "Great Grizzly Bear Hunt in Santa Paula Backcountry Reaps State Flag Icon, Tall Tales," *Ventura County Star,* Aug. 9, 2014.

32. Francesco di Castri and Harold A. Mooney, eds., *Mediterranean Type Ecosystems: Origin and Structure* (Berlin: Springer, 1973); Anderson, Barbour, and Whitworth, "World of Balance and Plenty"; Peter R. Dallman, *Plant Life in the World's Mediterranean Climates: California, Chile, South Africa, Australia, and the Mediterranean Basin* (Berkeley: University of California Press, and Sacramento: California Native Plant Society, 1998); Davis, *Ecology of Fear,* 12–13, 19; Dolph Schluter and Matthew W. Pennell, "Speciation Gradients and the Distribution of Biodiversity," *Nature* 546 (2017): 48–55; Martin L. Cody, "Diversity, Rarity, and Conservation in Mediterranean-Climate Regions," in *CB: The Science of Scarcity and Diversity,* ed. Michael E. Soulé (Sunderland, Mass.: Sinauer, 1986); Norman Myers et al., "Biodiversity Hotspots for Conservation Priorities," *Nature* 403 (2000): 853–858; Xiongwen Chen et al., "Spatial Structure of Multispecies Distributions in Southern California, USA," *BC* 124 (2005): 169–175; Schiffman, "Los Angeles Prairie"; Tershy et al., "Biodiversity"; C. R. Margules et al., "Systematic Conservation Planning,"

Nature 405 (2000): 243–253; Thomas M. Brooks et al., "Global Biodiversity Conservation Priorities," *Science* 313 (2006): 58–61; G. F. Midgley et al., "Developing Regional and Species-Level Assessments of Climate Change Impacts on Biodiversity in the Cape Floristic Region," *BC* 112 (2003): 87–97; Mirijam Gaertner et al., "Impacts of Alien Plant Invasion on Species Richness in Mediterranean-Type Ecosystems: A Meta-Analysis," *Progress in Physical Geography* 33 (2009): 319–338; Osvaldo E. Sala et al., "Global Biodiversity Scenarios for the Year 2100," *Science* 287 (2000): 1770–1774; David L. Wagner, "Insect Declines in the Anthropocene," *Annual Review of Entomology* 65 (2020): https://doi.org/10/1146annurev-ento-011019-025151; Sebastian Seibold et al., "Arthropod Decline in Grasslands and Forests Is Associated with Landscape-Level Drivers," *Nature* 574 (2019): 671–674; Kenneth V. Rosenberg et al., "Decline of the North American Avifauna," *Science* 366 (2019): 120–124; Ed Yong, "The Quiet Disappearance of Birds in North America," *Atlantic,* Sept. 19, 2019; Gerardo Ceballos et al., "Accelerated Modern Human-Induced Species Losses: Entering the Sixth Mass Extinction," *Science Advances* 1 (2015): https://doi.org/10.1126/sciadv.1400253; Audrey L. Mayer et al., "How Landscape Ecology Informs Global Land-Change Science and Policy," *BioScience* 66 (2016): 458–469; Matt Davis, Søren Faurby, and Jens-Christian Svenning, "Mammal Diversity Will Take Millions of Years to Recover from the Current Biodiversity Crisis," *PNAS* 115 (2018): 11262–11267; Intergovernmental Panel on Biodiversity and Ecosystem Services, "Nature's Dangerous Decline 'Unprecedented,' Species Extinction Rates 'Accelerating,'" Media Release, May 6, 2019, https://www.ipbes.net/news/Media-Release-Global-Assessment; Eyal Shochat et al., "From Patterns to Emerging Processes in Mechanistic Urban Ecology," *Trends in Ecology and Evolution* 21 (2006): 186–191; Sean A. Maxwell et al., "The Ravages of Guns, Nets and Bulldozers," *Nature* 536 (2016): 143–45.

Chapter 2. Essence of a California Gnatcatcher

1. Jonathan L. Atwood and David R. Bontrager, "California Gnatcatcher (*Polioptila californica*)," version 2.0, in *The Birds of North America,* ed. A. F. Poole and F. B. Gill (Cornell Lab of Ornithology, Ithaca, N.Y.), 2001, https://doi.org/10.2173/bna.574; Jonathan L. Atwood et al., "Distribution and Population Size of the California Gnatcatcher on the Palos Verdes Peninsula, 1993–1997," *WB* 29 (1998): 340–350.

2. Karen L. Miner, Adrian Wolf, and Robb Hirsch, "Use of Restored Coastal Sage Scrub Habitat by California Gnatcatchers in a Park Setting," *WB* 29 (1998): 439–446.

3. Atwood and Bontrager, "California Gnatcatcher."

4. Robert S. Woods, "Home Life of the Black-Tailed Gnatcatcher," *Condor* 23 (1921): 173–178.

5. Robert S. Woods, "Nesting of the Black-Tailed Gnatcatcher," *Condor* 30 (1928): 139–143 (quote on 143).

6. Atwood and Bontrager, "California Gnatcatcher." You can compare the calls of the California gnatcatcher to those of other gnatcatchers at the Cornell Lab of Ornithology's "All about Birds" webpages: https://www.allaboutbirds.org/.

7. Atwood and Bontrager, "California Gnatcatcher"; Sean A. Maxwell et al., "The Ravages of Guns, Nets and Bulldozers," *Nature* 536 (2016): 143–145.

8. Dan Froehlich, *Ageing North American Land Birds by Molt Limits and Plumage Criteria* (Bolinas, Calif.: Slate Creek Press, 2009); Atwood and Bontrager, "California Gnatcatcher."

9. Woods, "Home Life"; USFWS, "12-Month Finding on a Petition to Delist the Coastal California Gnatcatcher," 81 Fed. Reg. 59952, 59953 (Aug. 31, 2016).

10. Mark Avery, *A Message from Martha: The Extinction of the Passenger Pigeon and Its Relevance Today* (London: Bloomsbury, 2014); Barry Yeoman, "Why the Passenger Pigeon Went Extinct," *Audubon,* May–June 2014.

11. Michael A. Patten and Kurt F. Campbell, "Late Nesting of the California Gnatcatcher," *WB* 25 (1994): 110–111; J. Paul Galvin, "Breeding and Dispersal Biology of the California Gnatcatcher in Central Orange County," *WB* 29 (1998): 323–332; Mary A. Grishaver, Patrick J. Mock, and Kristine L. Preston, "Breeding Behavior of the California Gnatcatcher in Southwestern San Diego County, California," *WB* 29 (1998): 299–322; Keith W. Sockman, "Seasonal Variation in Nest Placement by the California Gnatcatcher," *Wilson Bulletin* 112 (2000): 498–504; Atwood and Bontrager, "California Gnatcatcher"; USFWS, *Coastal California Gnatcatcher 5-Year Review: Summary and Evaluation* (Carlsbad, Calif.: USFWS, Sept. 29, 2010), https://www.fws.gov/carlsbad/SpeciesStatusList/5YR/20100929_5YR_CAGN.pdf; Woods, "Home Life"; Woods, "Nesting"; David R. Bontrager, *Habitat Requirements, Home Range, and Breeding Biology of the California Gnatcatcher (Polioptila californica) in South Orange County, California* (Prepared for Santa Margarita Company, Rancho Santa Margarita, Calif., 1991; available from Wilson Ornithological Society, Museum of Zoology, University of Michigan, Ann Arbor); Michael A. Patten and John T. Rotenberry, "The Proximate Effects of Rainfall on Clutch Size of the California Gnatcatcher," *Condor* 101 (1999): 876–880.

12. Galvin, "Breeding and Dispersal Biology"; Grishaver, Mock, and Preston, "Breeding Behavior"; Patten and Rotenberry, "Proximate Effects of Rainfall"; David R. Bontrager et al., *1995 Breeding Biology of the California Gnatcatcher in the*

San Joaquin Hills, Orange County, California (Laguna Beach, Calif.: Superpark Project Rep., SP-95-04, available from Wilson Ornithological Society, Museum of Zoology, University of Michigan, Ann Arbor); Atwood and Bontrager, "California Gnatcatcher." Averages range from 1.6 to 4.4 fledglings per year depending on distance from coast and weather conditions: G. T. Braden, R. L. McKernan, and R. L. Powell, "Life History of *Polioptila californica* in Western Riverside County, CA" (Paper presented at the Symposium on the Biology of the California Gnatcatcher, University of California, Riverside, Sept. 15–16, 1995). Note that mortality numbers are based on this one study, on one study site, and do not appear in a published, peer-reviewed document.

13. Audrey L. Mayer, pers. obs.; Grishaver, Mock, and Preston, "Breeding Behavior"; Woods, "Home Life"; John D. Roach, "The Influence of Vegetation Structure and Arthropod Abundance on the Reproductive Success of California Black-Tailed Gnatcatchers, *Polioptila melanura californica*" (M.S. thesis, San Diego State University, 1989); Bontrager, *Habitat Requirements;* Jonathan L. Atwood, "California Gnatcatchers and Coastal Sage Scrub: The Biological Basis for Endangered Species Listing," in *Interface between Ecology and Land Development in California,* ed. Jon E. Keeley (Los Angeles: Southern California Academy of Science, 1993), 149–169; Galvin, "Breeding and Dispersal Biology"; Sockman, "Seasonal Variation in Nest Placement"; Woods, "Nesting"; Miner, Wolf, and Hirsch, "Use of Restored Coastal Sage Scrub Habitat"; Peter Famolaro and Jeff Newman, "Occurrence and Management Considerations of California Gnatcatchers along San Diego County Highways," *WB* 29 (1998): 447–452; Atwood and Bontrager, "California Gnatcatcher."

14. Woods, "Home Life"; Grishaver, Mock, and Preston, "Breeding Behavior"; Patten and Campbell, "Late Nesting"; Keith W. Sockman, "Nest Attendance by Male California Gnatcatchers," *Journal of Field Ornithology* 69 (1998): 95–102.

15. Atwood and Bontrager, "California Gnatcatcher"; E. Tattersall, "The California Black-Tailed Gnatcatcher" (M.S. thesis, University of California, Riverside, 1988); ERC Environmental and Energy Services Co., "Phase I Report, Amber Ridge California Gnatcatcher study" (Report prepared for Weingarten, Siegel, Fletcher Group, Inc., ERC Environmental and Energy Services, 5510 Morehouse Dr., San Diego, Calif., 92121, 1990); Patrick J. Mock et al., "Home Range Size and Habitat Preferences of the California Gnatcatcher in San Diego County" (Paper presented at American Ornithologists' Union and Cooper's Ornithological Society Joint Annual Meeting, Los Angeles, June 1990); Bontrager, *Habitat Requirements;* Jonathan L. Atwood, "A Maximum Estimate of the California Gnatcatcher's Population Size in the United States," *WB* 23 (1992): 1–9; Atwood, "California Gnatcatchers and Coastal

Sage Scrub"; H. Reşit Akçakaya and Jonathan L. Atwood, "A Habitat-Based Meta-population Model of the California Gnatcatcher," *CB* 11 (1997): 422–434; Galvin, "Breeding and Dispersal Biology"; Kristine L. Preston et al., "California Gnatcatcher Territorial Behavior," *WB* 29 (1998): 242–257.

16. Woods, "Nesting," 143.

17. This is a big assumption. See Daniel Rubinoff, "Evaluating the California Gnatcatcher as an Umbrella Species for Conservation of Southern California Coastal Sage Scrub," *CB* 15 (2001): 1374–1383.

18. For example, Woods, "Home Life."

19 Preston et al., "California Gnatcatcher Territorial Behavior." Small, isolated subpopulations are more likely to decline to zero than more connected subpopulations. Robert H. MacArthur and Edward O. Wilson, *Island Biogeography* (Princeton, N.J.: Princeton University Press, 1967); Mark Shaffer, "Minimum Viable Populations: Coping with Uncertainty," in *Viable Populations for Conservation,* ed. Michael E. Soulé (Cambridge: Cambridge University Press, 1987).

20. Eric A. Bailey and Patrick J. Mock, "Dispersal Capability of the California Gnatcatcher: A Landscape Analysis of Distribution Data," *WB* 29 (1998): 351–360; Galvin, "Breeding and Dispersal Biology"; Preston et al., "California Gnatcatcher Territorial Behavior."

21. W. T. Everett, P. Unitt, and A. M. Rea, "Investigations into the Status of the California Gnatcatcher on Point Loma, San Diego, California" (Report prepared for Natural Resources Management Branch, Southwest Division, Naval Facilities Engineering Command, San Diego, 1993, available from the Wilson Ornithological Society, Museum of Zoology, University of Michigan, Ann Arbor); Jonathan L. Atwood et al., "Population Dynamics, Dispersal and Demography of California Gnatcatchers and Cactus Wrens in Coastal Southern California (1997 progress report)" (Report, Manomet Observatory for Conservation Sciences, Manomet, Mass., 1998, available from Wilson Ornithological Society, Museum of Zoology, University of Michigan, Ann Arbor); Jonathan L. Atwood, David R. Bontrager, and Amy L. Gorospe, "Use of Refugia by California Gnatcatchers Displaced by Habitat Loss," *WB* 29 (1998): 406–412; Bailey and Mock, "Dispersal Capability"; Galvin, "Breeding and Dispersal Biology"; Atwood and Bontrager, "California Gnatcatcher"; Patrick J. Mock, "California Gnatcatcher (*Polioptila californica*)," in *The Coastal Scrub and Chaparral Bird Conservation Plan: A Strategy for Protecting and Managing Coastal Scrub and Chaparral Habitats and Associated Birds in California,* ed. John C. Lovio (California Partners in Flight, 2004), http://www.prbo.org/calpif/htmldocs/scrub.html; USFWS, "Revised Designation of Critical Habitat for the Coastal California Gnat-

catcher (*Polioptila californica californica*); Final Rule," 72 Fed. Reg. 72010 (Dec. 19, 2007); USFWS, *Coastal California Gnatcatcher 5-Year Review.*

22. Amy G. Vandergast et al., "Distinguishing Recent Dispersal from Historical Genetic Connectivity in the Coastal California Gnatcatcher," *Scientific Reports* 9 (2019): art1355, https://doi.org/10.1038/s41598-018-37712-2; USFWS, "Determination of Threatened Status for the Coastal California Gnatcatcher," 58 Fed. Reg. 16742 (Mar. 30, 1993); Glenn D. Sutherland et al., "Scaling of Natal Dispersal Distances in Terrestrial Birds and Mammals," *Conservation Ecology* 4 (2000): https://www.jstor.org/stable/26271738; Akçakaya and Atwood, "Habitat-Based Metapopulation Model"; Vandergast et al., "Distinguishing Recent Dispersal"; John C. Lovio, "The Effects of Habitat Fragmentation on the Breeding-Bird Assemblage in California Coastal Sage Scrub" (M.S. thesis, San Diego State University, 1996); Bailey and Mock, "Dispersal Capability."

23. Liam H. Davis, Robert L. McKernan, and James S. Burns, "History and Status of the California Gnatcatcher in San Bernardino County, California," *WB* 29 (1998): 361–365, 364.

24. Kathryn Dunn, "Claremont Brush Fire Burns Four Acres," *Claremont (Calif.) Courier,* May 19, 2017.

25. Spencer F. Baird, Thomas M. Brewer, and Robert Ridgway, *A History of North American Birds* (Boston: Little, Brown, 1874); Woods, "Home Life"; Jutta C. Burger et al., "Foraging Ecology of the California Gnatcatcher Deduced from Fecal Samples," *Oecologia* 120 (1999): 304–310; Woods, "Home Life"; Atwood and Bontrager, "California Gnatcatcher."

26. Steven R. Beissinger and David R. Osborne, "Effects of Urbanization on Avian Community Organization," *Condor* 84 (1982): 75–83; Ian MacGregor-Fors, Lorena Morales-Pérez, and Jorge E. Schondube, "From Forests to Cities: Effects of Urbanization on Tropical Birds," in *Urban Bird Ecology and Conservation,* ed. Christopher A. Lepczyk and Paige S. Warren (Berkeley: University of California Press, 2012), 33–48; Amanda D. Rodewald, "Evaluating Factors That Influence Avian Community Response to Urbanization," in Lepczyk and Warren, *Urban Bird Ecology and Conservation,* 71–92; Mark A. Goddard, Karen Ikin, and Susannah B. Lerman, "Ecological and Social Factors Determining the Diversity of Birds in Residential Yards and Gardens," in *Ecology and Conservation of Birds in Urban Environments,* ed. Enrique Murgui and Marcus Hedblom (Cham, Switzerland: Springer, 2017), 371–398.

27. Atwood and Bontrager, "California Gnatcatcher"; Clark S. Winchell and Paul F. Doherty Jr., "Using California Gnatcatcher to Test Underlying Models in Habitat Conservation Plans," *Journal of Wildlife Management* 72 (2008): 1322–1327;

Jennifer R. Vaughan, "Local Geographies of the Coastal Cactus Wren and the Coastal California Gnatcatcher on Marine Corps Base Camp Pendleton, California" (M.S. thesis, San Diego State University, 2010). 28. Frank W. Davis, Peter A. Stine, and David M. Stoms, "Distribution and Conservation Status of Coastal Sage Scrub in Southwestern California," *Journal of Vegetation Science* 5 (1994): 743–756; J. C. Burger et al., "Restoring Arthropod Communities in Coastal Sage Scrub," *CB* 17 (2003): 460–467; Walter E. Westman, "Factors Influencing the Distribution of Species of Californian Coastal Sage Scrub," *Ecology* 62 (1981): 439–455; Patrick J. Mock, "Energetic Constraints to the Distribution and Abundance of the California Gnatcatcher," *WB* 29 (1998): 413–420; Walter H. Muller and Cornelius H. Muller, "Volatile Growth Inhibitors Produced by Salvia Species," *Bulletin of the Torrey Botanical Club* 91 (1964): 327–330; Cornelius H. Muller and Roger del Moral, "Soil Toxicity Induced by Terpenes from *Salvia leucophylla*," *Bulletin of the Torrey Botanical Club* 93 (1966): 130–137; Bennet J. Tyson, William A. Dement, and Harold A. Mooney, "Volatilization of Terpenes from *Salvia mellifera*," *Nature* 252 (1974): 119–120; Bontrager, *Habitat Requirements;* Bontrager et al., *1995 Breeding Biology;* Sweetwater Environmental Biologists, "Orange County Parks Coastal California Gnatcatcher and San Diego Cactus Wren Survey Report" (Report prepared for Orange County Environmental Management Agency, 1994, available from Wilson Ornithological Society, Museum of Zoology, University of Michigan, Ann Arbor); Kenneth L. Weaver, "Coastal Sage Scrub Variations of San Diego County and Their Influence on the Distribution of the California Gnatcatcher," *WB* 29 (1998): 392–405; Mary K. Chase et al., "Single Species as Indicators of Species Richness and Composition in California Coastal Sage Scrub Birds and Small Mammals," *CB* 14 (2000): 474–487; Rubinoff, "Evaluating the California Gnatcatcher."

29. Woods, "Home Life," 173; Jonathan L. Atwood, "Status Review of the California Gnatcatcher (*Polioptila californica*)" (Report, Manomet Bird Observatory, PO Box 1770, Manomet, Mass., 02345, 1990); Atwood and Bontrager, "California Gnatcatcher"; 72 Fed. Reg. 72010; 81 Fed. Reg. 59952; Jonathan L. Atwood and Jeffrey S. Bolsinger, "Elevational Distribution of California Gnatcatchers in the United States," *Journal of Field Ornithology* 63 (1992): 159–168.

30. Atwood and Bolsinger, "Elevational Distribution"; Patrick J. Mock and Douglas T. Bolger, "Ecology of the California Gnatcatcher at Rancho San Diego" (Technical appendix to the Rancho San Diego Habitat Conservation Plan, prepared by Ogden Environmental and Energy Services for Home Capital Development Corp., 1992, available from Wilson Ornithological Society, Museum of Zoology, University of Michigan, Ann Arbor); Patrick J. Mock, "Population Viability Analy-

sis of the California Gnatcatcher within the MSCP Study Area" (Report prepared by Ogden Environmental and Energy Services for the City of San Diego Clean Water Program, 1993, available from Wilson Ornithological Society, Museum of Zoology, University of Michigan, Ann Arbor); Heather L. Hulton VanTassel et al., "Environmental Change, Shifting Distributions, and Habitat Conservation Plans: A Case Study of the California Gnatcatcher," *Ecology and Evolution* 7 (2017): 10326–10338. A regionwide survey was conducted in 2016; as of May 2020 results were not yet published, but a map of surveyed and occupied points is available here: https://sdmmp .com/view_project.php?sdid=SDID_201612021615.5. Kenneth L. Weaver, "A New Site of Sympatry of the California and Black-Tailed Gnatcatchers in the United States," *WB* 29 (1998): 476–479; Joseph Grinnell, "Midwinter Birds at Palm Springs, California," *Condor* 6 (1904): 40–45; Mock, "Energetic Constraints."

31. Grinnell, "Birds of the Pacific Slope"; Robert S. Woods, "Black-Tailed Gnatcatcher," in *Life Histories of North American Thrushes, Kinglets, and Their Allies,* ed. Arthur Cleveland Bent, *Bulletin of the United States National Museum* 196 (1949), 374–381; Atwood, "California Gnatcatchers and Coastal Sage Scrub"; Weaver, "Coastal Sage Scrub Variations of San Diego County"; Atwood and Bontrager, "California Gnatcatcher"; Mock, "California Gnatcatcher"; Laura Yetter, "Landscape Pattern a Determinate of Coastal California Gnatcatcher (*Polioptila californica californica*) Occupation" (M.S. thesis, California State University, Northridge, 2012); Akçakaya and Atwood, "Habitat-Based Metapopulation Model"; David B. Lindenmayer and Joern Fischer, *Habitat Fragmentation and Landscape Change: An Ecological and Conservation Synthesis* (Washington, D.C.: Island Press, 2006).

32. Bontrager, *Habitat Requirements;* Atwood and Bolsinger, "Elevational Distribution"; Atwood, "California Gnatcatchers and Coastal Sage Scrub"; Jonathan L. Atwood, "Analysis of Edge Effects on California Gnatcatcher Reproductive Success" (Report prepared for the City of Rancho Palos Verdes, 1998); Richard A. Erickson and Karen L. Miner, "Six Years of Synchronous California Gnatcatcher Fluctuations at Two Locations in Coastal Orange County, California," *WB* 29 (1998): 333–339; Mock, "Energetic Constraints"; Weaver, "Coastal Sage Scrub Variations of San Diego County"; Peter A. Bowler, "Ecological Restoration of Coastal Sage Scrub and Its Potential Role in Habitat Conservation Plans," *Environmental Management* 26 (2000): S85–S96; Mock, "California Gnatcatcher"; Yetter, "Landscape Pattern."

Chapter 3. Population Trends and Current Threats

1. Joseph Grinnell, "Birds of the Pacific Slope of Los Angeles County," *Pasadena Academy of Sciences* 11 (1898): 50.

2. Joseph Grinnell and Alden H. Miller, "The Distribution of the Birds of California," *Pacific Coast Avifauna*, no. 27 (Berkeley, Calif.: Cooper Ornithological Society, 1944), 369; Robert L. Pyle and Arnold Small, *Annotated Field List, Birds of Southern California* (Los Angeles: Los Angeles Audubon Society, 1961); R. Guy McCaskie and Eleanor A. Pugh, "Nesting Season: Southern Pacific Coast Region," *Audubon Field Notes* 18 (1964): 534–536; USFWS, "Endangered and Threatened Wildlife and Plants; Proposed Rule to List the Coastal California Gnatcatcher as Endangered," 56 Fed. Reg. 47053 (Sept. 17, 1991).

3. Jonathan L. Atwood, "The United States Distribution of the California Black-Tailed Gnatcatcher," *WB* 11 (1980): 65–78; Jonathan L. Atwood, "A Maximum Estimate of the California Gnatcatcher's Population Size in the United States," *WB* 23 (1992): 1–9.

4. Atwood, "Maximum Estimate."

5. USFWS, "Reinitiation of Formal Consultation on Implementation of the Special Rule for the Coastal California Gnatcatcher (1-6-93-FW-37RI)" (Report prepared for USFWS, California State Supervisor, Sacramento, 1996), 33; Jonathan L. Atwood and David R. Bontrager, "California Gnatcatcher (*Polioptila californica*)," version 2.0, in *The Birds of North America*, ed. A. F. Poole and F. B. Gill (Cornell Lab of Ornithology, Ithaca, N.Y.), 2001, https://doi.org/10.2173/bna.574.

6. The lack of rangewide monitoring programs is quite common among listed species; Daniel M. Evans et al., "Species Recovery in the United States: Increasing the Effectiveness of the Endangered Species Act," *Issues in Ecology* 20 (2016): 1–28; USFWS, *Coastal California Gnatcatcher 5-Year Review: Summary and Evaluation* (Carlsbad, Calif.: USFWS, Sept. 29, 2010), https://www.fws.gov/carlsbad/SpeciesSta tusList/5YR/20100929_5YR_CAGN.pdf; Louis Sahagun, "Developers Seek End to Federal Protections for California Gnatcatcher," *LAT*, June 29, 2014; Louis Sahagun, "Take the Gnatcatcher Off the Endangered List? Here's Why Wildlife Officials Say No," *LAT*, Aug. 31, 2016.

7. USFWS, "12-Month Finding on a Petition to Delist the Coastal California Gnatcatcher," 81 Fed. Reg. 59952, 59953 (Aug. 31, 2016).

8. Mission and Vision, San Diego Management and Monitoring Program, https://sdmmp.com/about.php#history; California Gnatcatcher South Coastal Monitoring Program, San Diego Management and Monitoring Program, https://sdmmp.com/view_project.php?sdid=SDID_201612021615.5; Alexandra Houston, Barbara Kus, and Kris Preston, "CAGN Occupancy Summary Sept. 2017 Presentation PPT," https://sdmmp.com/view_article.php?cid=CID_sarah.mccutcheon@aecom.com_59 d2bfa3a077d; Natural Communities Coalition, *Nature Reserve of Orange County,*

County of Orange Central/Coastal NCCP/HCP, 2017 Annual Report, https://occonser
vation.org/wp-content/uploads/2018/11/2017-Annual-Report-v2-reduced-size.pdf;
Barbara Kus, USGS, pers. comm., May 15, 2019.

9. North American Breeding Bird Survey data available at https://www.pwrc.usgs
.gov/bbs/RawData; Keith L. Pardieck et al., "North American Breeding Bird Survey
Dataset, 1966–2017, version 2017.0" (U.S. Geological Survey, Patuxent Wildlife Research Center, 2018). National Audubon Society, The Christmas Bird Count Historical Results, 2010; Christmas Bird Count Data are provided by the National Audubon
Society and through the generous efforts of Bird Studies Canada and countless vol-
unteers across the Western Hemisphere; CBC Data available at http://netapp.audu
bon.org/cbcobservation/. Winchell and Doherty, "Using California Gnatcatcher";
William B. Miller and Clark S. Winchell, "A Comparison of Point-Count and Area-
Search Surveys for Monitoring Site Occupancy of the Coastal California Gnatcatcher
(*Polioptila californica californica*)," *Condor* 118 (2016): 329–337; USFWS, "Determination of Threatened Status for the Coastal California Gnatcatcher," 58 Fed. Reg.
16742, 16747–16748 (Mar. 30, 1993).

10. Jonathan L. Atwood et al., "Distribution and Population Size of the California Gnatcatcher on the Palos Verdes Peninsula, 1993–1997," *WB* 29 (1998): 340–350;
Richard A. Erickson and Karen L. Miner, "Six Years of Synchronous California Gnatcatcher Fluctuations at Two Locations in Coastal Orange County, California," *WB* 29
(1998): 333–339; USFWS, *Coastal California Gnatcatcher 5-Year Review.*

11. eBird, *eBird: An Online Database of Bird Distribution and Abundance* (web application), https://ebird.org/home.

12. Marla Cone, "Gnatcatcher Study Rebuts Finding That It's Imperiled," *LAT,*
July 24, 1991.

13. Jonathan L. Atwood, "California Gnatcatchers and Coastal Sage Scrub: The
Biological Basis for Endangered Species Listing," in *Interface between Ecology and
Land Development in California,* ed. Jon E. Keeley (Los Angeles: Southern Califor-
nia Academy of Science, 1993); Patrick J. Mock, "Energetic Constraints to the Dis-
tribution and Abundance of the California Gnatcatcher," *WB* 29 (1998): 413–420;
USFWS, "Proposed Determination of Critical Habitat for the Coastal California
Gnatcatcher," 65 Fed. Reg. 5946 (Feb. 7, 2000); Atwood and Bontrager, "California
Gnatcatcher"; USFWS, "Revised Designation of Critical Habitat for the Coastal
California Gnatcatcher (*Polioptila californica californica*); Final Rule," 72 Fed. Reg.
72010 (Dec. 19, 2007); Patrick J. Mock, "California Gnatcatcher (*Polioptila californica*)," in *The Coastal Scrub and Chaparral Bird Conservation Plan: A Strategy for Protecting and Managing Coastal Scrub and Chaparral Habitats and Associated Birds in*

California, ed. John C. Lovio (California Partners in Flight, 2004, http://www.prbo .org/calpif/htmldocs/scrub.html); Daniel S. Cooper, Jennifer Mongolo, and Chris Dellith, "Status of the California Gnatcatcher at the Northern End of Its Range," *WB* 48 (2017): 124–140; Amy G. Vandergast et al., "Distinguishing Recent Dispersal from Historical Genetic Connectivity in the Coastal California Gnatcatcher," *Scientific Reports* 9 (2019): art1355, https://doi.org/10.1038/s41598-018-37712-2.

14. Atwood et al., "Distribution and Population Size."

15. H. Reşit Akçakaya and Jonathan L. Atwood, "A Habitat-Based Metapopulation Model of the California Gnatcatcher," *CB* 11 (1997): 422–434; Atwood and Bontrager, "California Gnatcatcher."

16. Atwood, "United States Distribution"; Liam H. Davis, Robert L. McKernan, and James S. Burns, "History and Status of the California Gnatcatcher in San Bernardino County, California," *WB* 29 (1998): 361–365; Amy G. Vandergast et al., "Are Hotspots of Evolutionary Potential Adequately Protected in Southern California?" *BC* 141 (2008): 1648–1664.

17. Atwood, "United States Distribution"; Atwood, "California Gnatcatchers and Coastal Sage Scrub"; Davis, McKernan, and Burns, "History and Status," 364; Atwood and Bontrager, "California Gnatcatcher."

18. 58 Fed. Reg. 16742; Marla Cone, "Builders, Tollway Agency Ask Judge to Block Gnatcatcher Decision," *LAT,* Feb. 25, 1993. The unpublished consultant's report is by H. Lee Jones at Michael Brandman Associates. Cone, "Gnatcatcher Study Rebuts Finding." It is possible that Cone reported 2.5 million *birds* in the 1991 article but 2.5 million *pairs* in the 1993 article. Either is equally improbable. Pacific Legal Foundation, "Flocks of California Gnatcatchers Need No Federal Protection" (undated blog post), https://pacificlegal.org/case/center-for-environmental-science-accu racy-reliability-et-al-v-u-s-department-of-interior-et-al/; 81 Fed. Reg. 59952, 59972.

19. J. B. Kirkpatrick and C. F. Hutchinson, "The Community Composition of California Coastal Sage Scrub," *Vegetatio* 35 (1977): 21–33; Walter E. Westman, "Factors Influencing the Distribution of Species of Californian Coastal Sage Scrub," *Ecology* 62 (1981): 439–455; J. F. O'Leary, "Coastal Sage Scrub: Threats and Current Status," *Fremontia* 23 (1995): 27–31; Jay E. Diffendorfer et al., "Developing Terrestrial, Multi-Taxon Indices of Biological Integrity: An Example from Coastal Sage Scrub," *BC* 140 (2007): 130–141; Heather L. Hulton VanTassel et al., "Environmental Change, Shifting Distributions, and Habitat Conservation Plans: A Case Study of the California Gnatcatcher," *Ecology and Evolution* 7 (2017): 10326–10338. Given that exotic grasses were introduced to the area as early as the 1500s by Spanish explorers and that Native Americans may have used fire to maintain grasslands over

coastal sage scrub, this 60–90 percent loss may be an overestimate, depending on the choice of baseline year (e.g., landscape dominated by indigenous tribes, Spanish arrival, mission times, ranches, etc.), as pointed out in Richard A. Minnich and Raymond J. Dezzani, "Historical Decline of Coastal Sage Scrub in the Riverside-Perris Plain, California," *WB* 29 (1998): 366–391; Jonathan L. Atwood, "Status Review of the California Gnatcatcher (*Polioptila californica*)" (unpublished report, Manomet Bird Observatory, PO Box 1770, Manomet, Mass., 02345, 1990); Douglas T. Bolger, Allison C. Alberts, and Michael E. Soulé, "Occurrence Patterns of Bird Species in Habitat Fragments: Sampling, Extinction, and Nested Species Subsets," *American Naturalist* 137 (1991): 155–166; Atwood, "Maximum Estimate," 7.

20. 81 Fed. Reg. 59952.

21. 81 Fed. Reg. 59952; David B. Lindenmayer and Joern Fischer, *Habitat Fragmentation and Landscape Change: An Ecological and Conservation Synthesis* (Washington, D.C.: Island Press, 2006).

22. 58 Fed. Reg. 16742, 16758–16759; Elsa E. Cleland, Jennifer L. Funk, and Edith B. Allen, "Coastal Sage Scrub," in *Ecosystems of California,* ed. Harold Mooney and Erika Zavaleta (Oakland: University of California Press, 2016), 429–448; 81 Fed. Reg. 59952, 59965.

23. 58 Fed. Reg. 16742; Atwood and Bontrager, "California Gnatcatcher"; Hulton VanTassel et al., "Environmental Change."

24. USFWS, *Coastal California Gnatcatcher 5-Year Review;* Lindenmayer and Fischer, *Habitat Fragmentation;* Clark S. Winchell and Paul F. Doherty Jr., "Effects of Habitat Quality and Wildfire on Occupancy Dynamics of Coastal California Gnatcatcher (*Polioptila californica californica*)," *Condor: Ornithological Applications* 116 (2014): 538–545.

25. Amy G. Vandergast et al., "Genetic Structure in the California Gnatcatcher in Coastal Southern California and Implications for Monitoring and Management" (Sacramento, Calif.: U.S. Department of the Interior, U.S. Geological Survey, 2014), https://nrm.dfg.ca.gov/FileHandler.ashx?DocumentID=97711; Vandergast et al., "Distinguishing Recent Dispersal"; Louis Sahagun, "Ecologists' Hopes Soar on the Wings of a Gnatcatcher," *LAT,* Dec. 5, 2006; 81 Fed. Reg. 59952.

26. Atwood, "California Gnatcatchers and Coastal Sage Scrub"; Mock, "Energetic Constraints"; Cooper, Mongolo, and Dellith, "Status."

27. Minnich and Dezzani, "Historical Decline"; Evan Meyer, Jorge Simancas, and Nick Jensen, "Conservation at California's Edge," *Fremontia* 44 (2016): 8–15; 81 Fed. Reg. 59952; Jason Henry, "City of Industry Offers $100 Million for Tres Hermanos' 2,500 Acres of Near Pristine Land," *San Gabriel Valley Tribune,* Jan. 13, 2017; Jason

Henry, "Why Does City of Industry Want Thousands of Acres of Ranchland in Chino Hills and Diamond Bar? Here's Their Plan," *San Gabriel Valley Tribune,* May 26, 2017; Ivan Penn and Ryan Menezes, "Californians Are Paying Billions for Power They Don't Need," *LAT,* Feb. 5, 2017; Jason Henry, "With Death of Solar Farm Project, Warring Cities Explore Shared Control of Tres Hermanos Ranch," *Pasadena Star News,* Aug. 19, 2018; Jim Gallagher, as quoted in Henry, "With Death of Solar Farm."

28. 81 Fed. Reg. 59952.

29. 81 Fed. Reg. 59952; David F. DeSante and Geoffrey R. Geupel, "Landbird Productivity in Central Coastal California: The Relationship to Annual Rainfall and a Reproductive Failure in 1986," *Condor* 89 (1987): 636–653; Mock, "Energetic Constraints"; Kristine L. Preston et al., "California Gnatcatcher Territorial Behavior," *WB* 29 (1998): 242–257; Atwood et al., "Distribution and Population Size"; Erickson and Miner, "Six Years of Synchronous California Gnatcatcher Fluctuations"; Richard T. Holmes, Thomas W. Sherry, and Franklin W. Sturges, "Bird Community Dynamics in a Temperate Deciduous Forest: Long-Term Trends at Hubbard Brook," *Ecological Monographs* 56 (1986): 201–220; Mary A. Grishaver, Patrick J. Mock, and Kristine L. Preston, "Breeding Behavior of the California Gnatcatcher in Southwestern San Diego County, California," *WB* 29 (1998): 299–322.

30. Giovanni Rapacciuolo et al., "Beyond a Warming Fingerprint: Individualistic Biogeographic Responses to Heterogeneous Climate Change in California," *Global Change Biology* 20 (2014): 2841–2855; 81 Fed. Reg. 59952; David W. Pierce et al., "Probabilistic Estimates of Future Changes in California Temperature and Precipitation Using Statistical and Dynamical Downscaling," *Climate Dynamics* 40 (2013): 839–856; Rapacciuolo et al., "Beyond a Warming Fingerprint"; Shelby Grad, "Most of California Is Out of the Drought," *LAT,* Feb. 23, 2017.

31. 81 Fed. Reg. 59952; Mock, "Energetic Constraints"; Zachary Principe et al., *50-Year Climate Scenarios and Plant Species Distribution Forecasts for Setting Conservation Priorities in Southwestern California* (San Francisco: Nature Conservancy of California, 2013); Alejandro E. Camacho, Elizabeth Taylor, and Melissa Kelly, "Lessons from Area-Wide, Multi-Agency Habitat Conservation Plans in California," *ELR* 46 (2016): 10222–10248.

32. Anne E. Kelly and Michael L. Goulden, "Rapid Shifts in Plant Distribution with Recent Climate Change," *PNAS* 105 (2008): 11823–11826; Frank W. Davis and Elizabeth A. Chornesky, "Adapting to Climate Change in California," *Bulletin of the Atomic Scientists* 70 (2014): 62–73; 81 Fed. Reg. 59952; Principe et al., *50-Year Climate Scenarios;* Zachary A. Holden et al., "Decreasing Fire Season Precipitation Increased Recent Western U.S. Forest Wildfire Activity," *PNAS* 115 (2018): E8349–E8357.

33. Mike Davis, *Ecology of Fear: Los Angeles and the Imagination of Disaster* (New York: Vintage Books, 1998), 95–112; Genie M. Fleming, James E. Diffendorfer, and Paul H. Zedler, "The Relative Importance of Disturbance and Exotic-Plant Abundance in California Coastal Sage Scrub," *Ecological Applications* 19 (2009): 2210–2227; Janet Franklin et al., "Altered Fire Regimes Affect Landscape Patterns of Plant Succession in the Foothills and Mountains of Southern California," *Ecosystems* 8 (2005): 885–898; Diffendorfer et al., "Developing Terrestrial, Multi-Taxon Indices"; Alexandra D. Syphard et al., "Human Influence on California Fire Regimes," *Ecological Applications* 17 (2007): 1388–1402; Diane E. Pataki et al., "Urban Ecosystems," in Mooney and Zavaleta, *Ecosystems of California,* 885–898; Peter R. Dallman, *Plant Life in the World's Mediterranean Climates: California, Chile, South Africa, Australia, and the Mediterranean Basin* (Berkeley: University of California Press, and Sacramento: California Native Plant Society, 1998); Davis, *Ecology of Fear,* 145.

34. Clark S. Winchell and Paul F. Doherty Jr., "Using California Gnatcatcher to Test Underlying Models in Habitat Conservation Plans," *Journal of Wildlife Management* 72 (2008): 1322–1327; Elizabeth F. van Mantgem, Jon E. Keeley, and Marti Witter, "Faunal Responses to Fire in Chaparral and Sage Scrub in California, USA," *Fire Ecology* 11 (2015): 128–148; 81 Fed. Reg. 59952; Jonathan L. Atwood, David R. Bontrager, and Amy L. Gorospe, "Use of Refugia by California Gnatcatchers Displaced by Habitat Loss," *WB* 29 (1998): 406–412; Audrey L. Mayer and William O. Wirtz II, "Effects of Fire on the Ecology of the California Gnatcatcher (*Polioptila californica*), and Associated Bird Species, in the Coastal Sage Scrub Community of Southern California," in *Brushfires in California Wildlands: Ecology and Resource Management,* ed. Jon E. Keeley and Tom Scott (Fairfield, Wash.: International Association of Wildland Fire, 1995), 70–77; Jan L. Beyers and William O. Wirtz II, "Vegetation Characteristics of Coastal Sage Scrub Sites Used by California Gnatcatchers: Implications for Management in a Fire-Prone Ecosystem," in *Fire Effects on Rare and Endangered Species and Habitats,* Proceedings of a Conference Held at Coeur d'Alene, Idaho, Nov. 13–16, 1995 (Fairfield, Wash.: International Association of Wildland Fire, 1997), 81–89; Peter A. Bowler, "Ecological Restoration of Coastal Sage Scrub and Its Potential Role in Habitat Conservation Plans," *Environmental Management* 26 (2000): S85–S96; USFWS, *Coastal California Gnatcatcher 5-Year Review;* 81 Fed. Reg. 59952; Diffendorfer et al., "Developing Terrestrial, Multi-Taxon Indices"; Fleming, Diffendorfer, and Zedler, "Relative Importance."

35. Jon E. Keeley and C. J. Fotheringham, "Historical Fire Regime in Southern California Shrublands," *CB* 15 (2001): 1536–1548; USFWS, *Coastal California Gnatcatcher 5-Year Review;* 81 Fed. Reg. 59952; Natural Communities Coalition, *Nature*

Reserve of Orange County, County of Orange Central/Coastal/NCCP/HCP, 2015 Annual Report, https://occonservation.org/wp-content/uploads/mdocs/2015annualreport.pdf.

36. Minnich and Dezzani, "Historical Decline"; Fleming, Diffendorfer, and Zedler, "Relative Importance"; 81 Fed. Reg. 59952; Wallace Kaufman and Syl Ramsey Kaufman, *Invasive Plants: Guide to Identification and the Impacts and Control of North American Species,* 2nd ed. (Mechanicsburg, Pa.: Stackpole Books, 2012), 12.

37. Atwood and Bontrager, "California Gnatcatcher"; Andrew F. Bennett and Denis A. Saunders, "Habitat Fragmentation and Landscape Change," in *CB for All,* ed. Navjot S. Sodhi and Paul R. Ehrlich (New York: Oxford University Press, 2010), 88–106; 81 Fed. Reg. 59952.

38. Peter E. Lowther, "Brown-Headed Cowbird (*Molothrus ater*)," version 1.0, in *Birds of the World,* ed. A. F. Poole and F. B. Gill (Ithaca, N.Y.: Cornell Lab of Ornithology, 2020), https://doi.org/10.2173/bow.bnhcow.01; Atwood and Bontrager, "California Gnatcatcher"; 81 Fed. Reg. 59952; Stephen I. Rothstein and Scott K. Robinson, "Conservation and Coevolutionary Implications of Brood Parasitism by Cowbirds," *Trends in Ecology and Evolution* 9 (1994): 162–164; Peter Arcese, James N. M. Smith, and Margret I. Hatch, "Nest Predation by Cowbirds and Its Consequences for Passerine Demography," *PNAS* 93 (1996): 4608–4611; Sharon Levy, "Native Incursions: Avian Range Expansions Imperil Threatened Species," *BioScience* 54 (2004): 94–98; Stephen A. Laymon, "Brown-Headed Cowbirds in California: Historical Perspectives and Management Opportunities in Riparian Habitats," *WB* 18 (1987): 63–70; Jean-Yves Barnagaud et al., "Dynamic Spatial Interactions between the Native Invader Brown-Headed Cowbird and Its Hosts," *Diversity and Distributions* 21 (2015): 511–522; S. Aki Hosoi and Stephen I. Rothstein, "Nest Desertion and Cowbird Parasitism: Evidence for Evolved Responses and Evolutionary Lag," *Animal Behaviour* 59 (2000): 823–840; USFWS, "90-Day Findings for Three Species; Notice of Petition Findings and Initiation of Status Reviews," 83 Fed. Reg. 30091 (June 27, 2018); USFWS, "Endangered and Threatened Wildlife and Plants; Removing the Kirtland's Warbler from the Federal List of Endangered and Threatened Wildlife; Final Rule," 84 Fed. Reg. 54436 (Oct. 19, 2019).

39. Gerald T. Braden, Robert L. McKernan, and Shawn M. Powell, "Effects of Nest Parasitism by the Brown-Headed Cowbird on Nesting Success of the California Gnatcatcher," *Condor* 99 (1997): 858–865; USFWS, *Coastal California Gnatcatcher 5-Year Review.*

40. Robert S. Woods, "Nesting of the Black-Tailed Gnatcatcher," *Condor* 30 (1928): 139–143; Braden, McKernan, and Powell, "Effects of Nest Parasitism"; USFWS, *Coastal California Gnatcatcher 5-Year Review,* 23; Michael A. Patten and

Douglas T. Bolger, "Variation in Top-Down Control of Avian Reproductive Success across a Fragmentation Gradient," *Oikos* 101 (2003): 479–488; Cleland, Funk, and Allen, "Coastal Sage Scrub."

41. Eric Mellink and Amadeo M. Rea, "Taxonomic Status of the California Gnat-catchers of Northwestern Baja California, Mexico," *WB* 25 (1994): 50–62; 81 Fed. Reg. 59952.

Chapter 4. California Gnatcatcher Taxonomy

1. Peter R. Grant and B. Rosemary Grant, "Hybridization of Bird Species," *Science* 256 (1992): 193–197; Chung-I. Wu, "The Genic View of the Process of Speciation," *Journal of Evolutionary Biology* 14 (2001): 851–865; Trevor D. Price and Michelle M. Bouvier, "The Evolution of F1 Postzygotic Incompatibilities in Birds," *Evolution* 56 (2002): 2083–2089; David Bickford et al., "Cryptic Species as a Window on Diversity and Conservation," *Trends in Ecology and Evolution* 22 (2007): 148–155; Frank E. Rheindt and Scott V. Edwards, "Genetic Introgression: An Integral but Neglected Component of Speciation in Birds," *Auk* 128 (2011): 620–632; John E. McCormack and James M. Maley, "Interpreting Negative Results with Taxonomic and Conservation Implications: Another Look at the Distinctness of Coastal California Gnatcatchers," *Auk* 132 (2015): 380–388; Bárbara R. N. Chaves et al., "Barcoding Neotropical Birds: Assessing the Impact of Nonmonophyly in a Highly Diverse Group," *Molecular Ecology Resources* 15 (2015): 921–931; Shannon J. Hackett et al., "A Phylogenomic Study of Birds Reveals Their Evolutionary History," *Science* 320 (2008): 1763–1768; Erich D. Jarvis et al., "Whole-Genome Analyses Resolve Early Branches in the Tree of Life of Modern Birds," *Science* 346 (2014): 1320–1331; Richard O. Prum et al., "A Comprehensive Phylogeny of Birds (Aves) Using Targeted Next-Generation DNA Sequencing," *Nature* 526 (2015): 569–573.

2. Eugene M. McCarthy, *Handbook of Avian Hybrids of the World* (New York: Oxford University Press, 2006), 80–83; Rheindt and Edwards, "Genetic Introgression"; Jennifer F. Lind-Riehl et al., "Hybridization, Agency Discretion, and Implementation of the U.S. Endangered Species Act," *CB* 30 (2016): 1288–1296.

3. Holly Doremus, "Listing Decisions under the Endangered Species Act: Why Better Science Isn't Always Better Policy," *Washington University Law Review* 75 (1997): 1029–1153; Teresa Woods and Steve Morey, "Uncertainty and the Endangered Species Act," *Indiana Law Journal* 83 (2008): 529–536; Jason Scott Johnston, "Introduction," in *Institutions and Incentives in Regulatory Science,* ed. Johnston (Plymouth, U.K.: Lexington Books, 2012), 1–14; Katrina M. Wyman, "Politics and Science in Endangered Species Act Listing," in Johnston, *Institutions and Incentives in Regula-*

tory Science, 99–130; Robert K. Wayne and H. Bradley Shaffer, "Hybridization and Endangered Species Protection in the Molecular Era," *Molecular Ecology* 25 (2016): 2680–2689; J. B. Ruhl, "Listing Endangered and Threatened Species," in *Endangered Species Act: Law, Policy, and Perspectives,* ed. Donald C. Baur and Wm. Robert Irvin, 2nd ed. (Chicago: American Bar Association, 2010), 16–39; Stephen M. Jackson et al., "The Wayward Dog: Is the Australian Native Dog or Dingo a Distinct Species?" *Zootaxa* 4317 (2017): 201–224; Euan G. Ritchie et al., "Species Definitions Shape Policy," *Science* 361 (2018): 1324.

4. Sue-Ellen Jacobs, Wesley Thomas, and Sabine Lang, eds., *Two-Spirit People: Native American Gender Identity, Sexuality, and Spirituality* (Champaign: University of Illinois Press, 1997); Joan Roughgarden, *Evolution's Rainbow: Diversity, Gender, and Sexuality in Nature and People* (Berkeley: University of California Press, 2004); Benoit Mandelbrot, *The Fractal Geometry of Nature* (New York: W. H. Freeman, 1982).

5. Matthew A. Cronin, "Systematics, Taxonomy, and the Endangered Species Act: The Example of the California Gnatcatcher (*Polioptila californica*)," *Wildlife Society Bulletin* 25 (1997): 661–666; Robert M. Zink et al., "Genetics, Taxonomy, and Conservation of the Threatened California Gnatcatcher," *CB* 14 (2000): 1394–1405; McCormack and Maley, "Interpreting Negative Results." Although Zink and his colleagues don't support subspecies in birds (and particularly not for the California gnatcatcher), they recently argued that there are so many cryptic species of birds (discoverable only using genetic methods) that the "real" number of bird species recognized should be double what it is now: George F. Barrowclough et al., "How Many Kinds of Birds Are There and Why Does It Matter?" *PLoS ONE* 11 (2016): https://doi.org/10.1371/journal.pone.0166307; R. Ruggles, "UNL Professor Argues There Are Twice as Many Bird Species as Believed, Rankling Some in Ornithology Community," *Omaha (Nebr.) World-Herald,* Jan. 3, 2017.

6. Th. A. Dobzhansky, "Critique of the Species Concept in Biology," *Philosophy of Science* 2 (1935): 344–355; Ernst Mayr, *Systematics and the Origin of Species* (New York: Columbia University Press, 1942).

7. J. Albert C. Uy, Darren E. Irwin, and Michael S. Webster, "Behavioral Isolation and Incipient Speciation in Birds," *Annual Review of Ecology, Evolution, and Systematics* 49 (2018): 1–24; Leigh Van Valen, "Ecological Species, Multispecies, and Oaks," *Taxon* 25 (1976): 233–239; R. J. Gutiérrez et al., "The Invasion of Barred Owls and Its Potential Effect on the Spotted Owl: A Conservation Conundrum," *Biological Invasions* 9 (2007): 181–196.

8. For descriptions of other species concepts, such as the "mitonuclear compat-

ibility" and "inclusive" species concepts, and a discussion of the problem of assigning firm boundaries to a fuzzy and moving target, see Frank E. Zachos, "(New) Species Concepts, Species Delimitation and the Inherent Limitations of Taxonomy," *Journal of Genetics* 97 (2018): 811–815. Post hoc choices of species concept fitting reproduction or other organismal traits risk elevating trivial differences to the species level, or "taxonomic inflation": Rob Roy Ramey II, "On the Origin of Specious Species," in *Institutions and Incentives in Regulatory Science,* ed. Jason Scott Johnston (Plymouth, U.K.: Lexington Books, 2012), 77–98; Cronin, "Systematics"; Stephen T. Garnett and Les Christidis, "Taxonomy Anarchy Hampers Conservation," *Nature* 546 (2017): 25–27; Quentin D. Wheeler and Norman I. Platnick, "The Phylogenetic Species Concept (*sensu* Wheeler & Platnick)," in *Species Concepts and Phylogenetic Theory: A Debate,* ed. Quentin Wheeler and Rudolf Meier (New York: Columbia University Press, 2000), 55–69; Michael A. Patten, "Subspecies and the Philosophy of Science," *Auk* 132 (2015): 481–485.

9. Katie Langin, "A Different Animal," *Science* 360 (2018): 590–592.

10. Lind-Riehl et al., "Hybridization"; Garnett and Christidis, "Taxonomy Anarchy." The U.S. Congress has attempted to clarify species concepts in the ESA: Doremus, "Listing Decisions"; Andrew Jenner, "The Factious, High-Drama World of Bird Taxonomy," *Atlantic,* Feb. 28, 2017; Doremus, "Listing Decisions"; SELS, *The Endangered Species Act* (Stanford, Calif.: Stanford University Press, 2001).

11. William Brewster, "On the Affinities of Certain *Polioptila* with a Description of a New Species," *Bulletin of the Nuttall Ornithological Club* 6 (1881): 101–107, quotation at 103; Eric Mellink and Amadeo M. Rea, "Taxonomic Status of the California Gnatcatchers of Northwestern Baja California, Mexico," *WB* 25 (1994): 50–62.

12. Bernie Tershy et al., "Biodiversity," in *Ecosystems of California,* ed. Harold Mooney and Erika Zavaleta (Oakland: University of California Press, 2016), 187–212; Joseph Grinnell and Alden H. Miller, "The Distribution of the Birds of California," *Pacific Coast Avifauna,* no. 27 (Berkeley, Calif.: Cooper Ornithological Society, 1944); Michael A. Patten et al., "Fifty Years since Grinnell and Miller: Where Is California Ornithology Headed?" *WB* 26 (1995): 54–64.

13. Joseph Grinnell, "Birds of the Pacific Slope of Los Angeles County," *Pasadena Academy of Sciences* 11 (1898): 1–52; Joseph Grinnell, "Midwinter Birds at Palm Springs, California," *Condor* 6 (1904): 45.

14. Joseph Grinnell, "A Critical Inspection of the Gnatcatchers of the Californias," *Proceedings of the California Academy of Sciences,* 4th ser., 15 (1926): 493–500; Mellink and Rea, "Taxonomic Status"; Zink et al., "Genetics, Taxonomy, and Conservation." In hindsight, towhees are a poor support for biogeographic arguments. *Pipilo*

fuscus was moved to an entirely different genus (*Melozone*) and split into two species based on genetics and geographic isolation: the canyon towhee (*Melozone fusca*) and the California towhee (*Melozone crissalis*). The taxonomy within the "brown towhee" species cluster remains contentious: Jeffrey M. DaCosta et al., "A Molecular Systematic Revision of Two Historically Problematic Songbird Clades: *Aimophila* and *Pipilo*," *Journal of Avian Biology* 40 (2009): 206-216.

15. Jonathan L. Atwood, "Speciation and Geographic Variation in Black-Tailed Gnatcatchers," *Ornithological Monographs* 42 (1988): 1-74; Mellink and Rea, "Taxonomic Status"; Jonathan L. Atwood and David R. Bontrager, "California Gnatcatcher (*Polioptila californica*)," version 2.0, in *The Birds of North America,* 2001, ed. A. F. Poole and F. B. Gill (Cornell Lab of Ornithology, Ithaca, N.Y.), https://doi.org /10.2173/bna.574.

16. Burt L. Monroe et al., "Thirty-Seventh Supplement to the American Ornithologists' Union Checklist of North American Birds," *Auk* 106 (1989): 532-538; McCormack and Maley, "Interpreting Negative Results"; Atwood, "Speciation and Geographic Variation"; Jonathan L. Atwood, "A Maximum Estimate of the California Gnatcatcher's Population Size in the United States," *WB* 23 (1992): 1-9.

17. Specifically, they used a rapidly evolving, noncoding region of mtDNA (mitochondrial DNA). Noncoding DNA are sections of chromosomes that do not direct the production of proteins, and these sections often change more rapidly than coding sections. Because these sections change faster, they can better reveal patterns of recent or closely spaced speciation events. Robert M. Zink and Rachelle C. Blackwell, "Molecular Systematics and Biogeography of Aridland Gnatcatchers (Genus *Polioptila*) and Evidence Supporting Species Status of the California Gnatcatcher (*Polioptila californica*)," *Molecular Phylogenetics and Evolution* 9 (1998): 26-32; Brewster, "Affinities"; John P. Hubbard, "Avian Evolution in the Aridlands of North America," in *The Living Bird, Twelfth Annual* (Ithaca, N.Y.: Cornell Laboratory of Ornithology, Apr. 1, 1974), 155-196; Atwood, "Speciation and Geographic Variation."

18. 50 C.F.R. § 424.02; Lind-Riehl et al., "Hybridization"; USFWS, "12-Month Finding on a Petition to Delist the Coastal California Gnatcatcher," 81 Fed. Reg. 59952 (Aug. 31, 2016). The term *evolutionarily significant unit* is also used, but mostly by the NMFS for protecting salmon runs. Langin, "Different Animal"; Justice Department, USFWS, and National Oceanic and Atmospheric Administration, "Policy Regarding the Recognition of Distinct Vertebrate Population Segments under the Endangered Species Act," 61 Fed. Reg. 4722 (Feb. 7, 1996); USFWS, "Endangered and Threatened Wildlife and Plants; Designation of Critical Habitat for the Coastal California Gnatcatcher (*Polioptila californica californica*) and Determination of Dis-

tinct Vertebrate Population Segment for the California Gnatcatcher (*Polioptila californica*); Proposed Rule," 68 Fed. Reg. 20228 (Apr. 24, 2003).

19. Atwood, "Speciation and Geographic Variation"; Richard C. Banks, "Review: Speciation and Geographic Variation in Black-Tailed Gnatcatchers," *Wilson Bulletin* 101 (1989): 360–362; Ned K. Johnson, "Review: Speciation and Geographic Variation in Black-Tailed Gnatcatchers," *Auk* 106 (1989): 347–349. The "*margaritae*" form discussed here is *P. c. margaritae,* now considered by some to be a subspecies of the California gnatcatcher. The "Cape form" is *P. c. abbreviata;* Jonathan L. Atwood, "Subspecies Limits and Geographic Patterns of Morphological Variation in California Gnatcatchers (*Polioptila californica*)," *Bulletin of Southern California Academy of Science* 90 (1991): 118–133. Rebecca Trounson and Mark I. Pinsky, "Gnatcatcher Taken off Threatened Species List," *LAT,* May 3, 1994; North American Classification and Nomenclature Committee, *American Ornithologists' Union Checklist,* 5th ed. (1957); Jonathan L. Atwood and Jeffrey S. Bolsinger, "Elevational Distribution of California Gnatcatchers in the United States," *Journal of Field Ornithology* 63 (1992): 159–168; Mellink and Rea, "Taxonomic Status"; 81 Fed. Reg. 59952; although see McCormack and Maley, "Interpreting Negative Results," which places it at 28°N; L. Lee Grismer, "Evolutionary Biogeography on Mexico's Baja California Peninsula: A Synthesis of Molecules and Historical Geology," *PNAS* 97 (2000): 14017–14018; Johan Lindell, Andre Ngo, and Robert W. Murphy, "Deep Genealogies and the Mid-Peninsular Seaway of Baja California," *Journal of Biogeography* 33 (2006): 1327–1331; Daniel I. Axelrod, "The Origin of Coastal Sage Vegetation, Alta and Baja California," *American Journal of Botany* 65 (1978): 1117–1131; Jon P. Rebman and Norman C. Roberts, *Baja California Plant Field Guide,* 3rd ed. (San Diego: San Diego National History Museum, 2012).

20. Grinnell, "Critical Inspection of Gnatcatchers"; A. J. Van Rossem, "Concerning Some *Polioptilae* of the West Coast of Middle America," *Auk* 48 (1931): 33–39; *American Ornithologists' Union Checklist,* 5th ed.; Atwood, "Speciation and Geographic Variation"; Atwood, "Subspecies Limits"; Mellink and Rea, "Taxonomic Status"; USFWS, "Endangered and Threatened Wildlife and Plants; Notice of Determination to Retain the Threatened Status for the Coastal California Gnatcatcher under the Endangered Species Act," 50 Fed. Reg. 15693 (Mar. 27, 1995); Zink and Blackwell, "Molecular Systematics"; John R. Skalski et al., "Type I Errors Linked to Faulty Statistical Analyses of Endangered Subspecies Classifications," *Journal of Agricultural, Biological, and Environmental Statistics* 13 (2008): 199–220.

21. Grinnell, "Critical Inspection of Gnatcatchers"; Van Rossem, "Concerning Some *Polioptilae*"; Allan R. Phillips, "¿Subspecies or especie olvidada? El caso de

Polioptila californica (Aves: Sylviidae)," (Resúmenes del 4to Congreso Nacional de Zoología, Ensenada, Baja California, [1980], 100); Sanford R. Wilbur, *Birds of Baja California* (Berkeley: University of California Press, 1987); Atwood, "Subspecies Limits"; USFWS, "Determination of Threatened Status for the Coastal California Gnatcatcher," 58 Fed. Reg. 16742 (Mar. 30, 1993); Mellink and Rea, "Taxonomic Status."

22. USFWS, ETWP, "Proposed Rule to List the Coastal California Gnatcatcher as Endangered," 56 Fed. Reg. 47053 (Sept. 17, 1991); Robert Reinhold, "Tiny Songbird Poses Big Test of U.S. Environmental Policy," *New York Times,* Mar. 16, 1993; Jonathan L. Atwood and Reed F. Noss, "Gnatcatchers and Development: A 'Train Wreck' Avoided?" *Illahee* 10 (1994): 123–130; 58 Fed. Reg. 16742–16753; McCormack and Maley, "Interpreting Negative Results."

23. 61 Fed. Reg. 4722; 58 Fed. Reg. 16742, 16750; 68 Fed. Reg. 20228.

24. Chris Clarke, "This Tiny Bird Scored a Win for Science," *KCET,* Sept. 21, 2016, https://www.kcet.org/redefine/this-tiny-bird-scored-a-win-for-science; USFWS, "Revised Designation of Critical Habitat for the Coastal California Gnatcatcher (*Polioptila californica californica*); Final Rule," 72 Fed. Reg. 72010 (Dec. 19, 2007); USFWS, "90-Day Findings on Two Petitions," 79 Fed. Reg. 78775 (Dec. 31, 2014); 81 Fed. Reg. 59952; see, e.g., Ctr. for Envtl. Sci., Accuracy & Reliability v. U.S. Dep't of Interior, No. CV 17-2313 (JDB), 2019 WL 2870131 (D.D.C. July 3, 2019); USFWS, "Endangered and Threatened Wildlife and Plants; Notice of Extension of Comment Period on Data Pertaining to the Subspecies Taxonomy of the California Gnatcatcher," 59 Fed. Reg. 38426 (July 28, 1994); see Endangered Species Comm. of Bldg. Indus. Ass'n of S. Cal. v. Babbitt, 852 F.Supp. 32, 43 (D.D.C. 1994), as amended on reconsideration (June 16, 1994); Atwood, "Subspecies Limits"; "Not Made Public as Required by Law": required by the Administrative Procedures Act, although Doremus ("Listing Decisions") argued that scientific tradition would also support the release of the raw data, e.g., if a peer reviewer requested it; Marla Cone, "Suit Targets Campaign to Protect Bird," *LAT,* Dec. 3, 1992; somewhat hypocritically (since Ramey stressed the need for better, more objective peer review for USFWS listing decisions); Ramey, "Specious Species," 87, cites eleven studies that retested Atwood's data once they were released publicly and finds that subspecies designations were not supported; ten of these studies remain "unpublished reports" without formal peer review; Trounson and Pinsky, "Gnatcatcher"; Atwood's statistical practices included extrapolating missing data and excluding outliers. Ramey ("Specious Species") points out that Atwood did not identify which data were estimates or identify the criteria for outliers: "In that deposition, [Atwood] recanted the reliability of key measurements,

admitted to substituting estimates for missing data, and told of destroying original copies of his data before he finished his dissertation and published the results." From "Crisis of Confidence: The Political Influence of the Bush Administration on Agency Science and Decision-Making," Oversight Hearing before the Committee on Natural Resources, U.S. House of Representatives, 110th Cong., 1st sess., July 31, 2007; see Endangered Species Comm., 852 F.Supp. at 43; Deborah Schoch, "Groups Drop Suit over Gnatcatcher's Threatened Status," *LAT,* Sept. 19, 1995.

25. Robert M. Zink, "The Role of Subspecies in Obscuring Biological Diversity and Misleading Conservation Policy," *PRSB* 271 (2004): 561–564; George Sangster, "The Application of Species Criteria in Avian Taxonomy and Its Implications for the Debate over Species Concepts," *Biological Reviews* 89 (2014): 199–214.

26. Zink and Blackwell, "Molecular Systematics"; Zink et al., "Genetics, Taxonomy, and Conservation."

27. John C. Avise and DeEtte Walker, "Pleistocene Phylogeographic Effects on Avian Populations and the Speciation Process," *PRSB* 265 (1998): 457–463. Although the "leading front" of a species sometimes has lower genetic diversity than in the core of its range, this is not always the case and thus should not be assumed: Amanda A. Pearce et al., "Genetic Variation during Range Expansion: Effects of Habitat Novelty and Hybridization," *PRSB* 284 (2017): https://doi.org/10.1098/rspb.2017.0007. Other factors can dictate genetic diversity at range extremes, such as the novelty of the habitat (less novel habitat would be expected to generate less genetic variability). Lower diversity is also expected when gene flow into these range fronts is reduced, which is increasingly the case for California gnatcatchers in the United States, because development along the U.S.-Mexico border limits dispersal.

28. Robert M. Zink et al., "Comparative Phylogeography of Some Aridland Bird Species," *Condor* 103 (2001): 1–10.

29. Skalski et al., "Type I Errors"; Mellink and Rea, "Taxonomic Status"; McCormack and Maley, "Interpreting Negative Results," 76 Fed. Reg. 66255 (Oct. 26, 2011).

30. Robert M. Zink, Sergei V. Drovetski, and Sievert Rohwer, "Phylogeographic Patterns in the Great Spotted Woodpecker *Dendrocopos major* across Eurasia," *Journal of Avian Biology* 33 (2002): 175–178; Zink, "Role of Subspecies"; Robert M. Zink, Jeffrey G. Groth, and Heran Vázquez-Miranda, "Phylogeography of the California Gnatcatcher (*Polioptila californica*) Using Multilocus DNA Sequences and Ecological Niche Modeling: Implications for Conservation," *Auk* 130 (2013): 449–458; Louis Sahagun, "Developers Seek End to Federal Protections for California Gnatcatcher," *LAT,* June 29, 2014; Patten, "Subspecies and the Philosophy of Science"; quote in McCormack and Maley, "Interpreting Negative Results," 383; J. V. Remsen Jr., "Pat-

tern, Process, and Rigor Meet Classification," *Auk* 122 (2005): 403–413; Kevin Winker et al., "Vainly Beating the Air: Species-Concept Debates Need Not Impede Progress in Science or Conservation," *Ornithological Monographs* 63 (2007): 30–44; Michael A. Patten, "Null Expectations in Subspecies Diagnosis," *Ornithological Monographs* 67 (2010): 35–41; R. L. Mayden, "A Hierarchy of Species Concepts: The Denouement in the Saga of the Species Problem," in *Species: The Units of Biodiversity,* ed. M. F. Claridge, H. A. Dawah, and M. R. Wilson (London: Chapman and Hall, 1997), 381–424; John C. Avise, "Cladists in Wonderland," *Evolution* 54 (2000): 1828–1832; Robert M. Zink et al., "Geographic Variation, Null Hypotheses, and Subspecies Limits in the California Gnatcatcher: A Response to McCormack and Maley," *Auk* 133 (2016): 59–68. Other funding sources for research supporting the no-subspecies side include two individuals, "S. Hills and D. Ely," who provided support for Cronin ("Systematics") and Zink et al. ("Genetics") and were revealed to be "Scott Hill and Doug Ely" in Skalski et al. ("Type I Errors"). Funding was for Zink et al., "Phylogeography," justified in Zink et al., "Geographic Variation," 64: "We understand that our failure to discover and disclose the fact that funding came from the Transportation Corridor Agency created a conflict of interest because Mr. Thornton has provided legal counsel in opposition to listing the California Gnatcatcher in 1994, has represented various land developers in Southern California that have contested the listing of the California Gnatcatcher, and was one of the 2 lawyers representing 7 plaintiffs in a petition to remove the California Gnatcatcher from the list of threatened species under the Endangered Species Act." Zink et al., "Geographic Variation," 65: "However, McCormack and Maley ('Interpreting Negative Results') imply that this conflict of interest might have influenced our interpretation of our data. We note that Mr. Thornton was not an author on the 2013 paper, and that neither Mr. Thornton nor the California state agency were given any opportunity to comment on our study design (e.g., choice of loci), analyses, interpretation of results, and design and preparation of the resulting manuscript. Furthermore, the University of Minnesota, through which the funds were disbursed, has strict rules that prohibit funding agencies from having any influence on the interpretation of results and subsequent publication of research. Thus, despite any apparent connections between 3rd parties with vested interests and our research, the conclusions in Zink et al. (2013) flowed exclusively from the data that we collected and analyzed to address the subspecific status of *P. c. californica.*" Doremus, "Listing Decisions"; "species cartels" from Ramey, "Specious Species"; Joe Roman, *Listed: Dispatches from America's Endangered Species Act* (Cambridge, Mass.: Harvard University Press, 2011), 170–173. See Atwood, "Maximum Estimate," 8 (acknowledgments); Atwood and Noss, "Gnatcatchers and Development"; and Jonathan L.

Atwood, David R. Bontrager, and Amy L. Gorospe, "Use of Refugia by California Gnatcatchers Displaced by Habitat Loss," *WB* 29 (1998): 406–412.

31. USFWS, "90-Day Finding on a Petition to Delist the Coastal California Gnatcatcher as Threatened," 76 Fed. Reg. 66255 (Oct. 26, 2011).

32. Zink et al., "Phylogeography"; 76 Fed. Reg. 66255; McCormack and Maley, "Interpreting Negative Results"; Zink et al., "Geographic Variation."

33. McCormack and Maley, "Interpreting Negative Results," 381; Robert M. Zink and George F. Barrowclough, "Mitochondrial DNA under Siege in Avian Phylogeography," *Molecular Ecology* 17 (2008): 2107–2121; George F. Barrowclough and Robert M. Zink, "Funds Enough, and Time: mtDNA, nuDNA and the Discovery of Divergence," *Molecular Ecology* 18, no. 14 (2009): 2934–2936; 81 Fed. Reg. 59952, 59960.

34. Sarah E. May et al., "Combining Genetic Structure and Ecological Niche Modeling to Establish Units of Conservation: A Case Study of an Imperiled Salamander," *BC* 144 (2011): 1441–1450; USFWS and NMFS, *Habitat Conservation Plans and Incidental Take Permit Processing Handbook* (2016), https://www.fws.gov/endangered/esa-library/pdf/HCP_Handbook.pdf; John E. McCormack, Amanda J. Zellmer, and L. Lacey Knowles, "Does Niche Divergence Accompany Allopatric Divergence in *Aphelocoma* Jays as Predicted under Ecological Speciation? Insights from Tests with Niche Models," *Evolution* 64 (2010): 1231–1244; Paige S. Warren et al., "Urban Ecology and Human Social Organisation," in *Urban Ecology,* ed. Kevin J. Gaston (Cambridge: Cambridge University Press, 2010); Zink et al., "Phylogeography"; 81 Fed. Reg. 59952; McCormack and Maley, "Interpreting Negative Results."

35. Sahagun, "Developers Seek"; Clarke, "This Tiny Bird"; McCormack and Maley, "Interpreting Negative Results"; 81 Fed. Reg. 59952.

36. Patten, "Subspecies and the Philosophy of Science"; 81 Fed. Reg. 59952; Mellink and Rea, "Taxonomic Status"; McCormack and Maley, "Interpreting Negative Results."

37. 58 Fed. Reg. 16742, 16750.

Chapter 5. The Gnatcatcher and the ESA

Note to epigraph: President Richard Nixon, State of the Union Address, February 15, 1973, https://www.presidency.ucsb.edu/documents/state-the-union-message-the-congress-natural-resources-and-the-environment.

1. Government attorney Donald Barry in 1990 called it "the pit bull of federal environmental statutes" in Robert D. Thornton, "Searching for Consensus and Predictability: Habitat Conservation Planning under the Endangered Species Act of

1973," *Environmental Law* 21 (1991): 605–656, quotation on 605; J. B. Ruhl, "Regional Habitat Conservation Planning under the Endangered Species Act: Pushing the Legal and Practical Limits of Species Protection," *Southwestern Law Journal* 44 (1991): 1393–1425; Timothy Beatley, "Reconciling Urban Growth and Endangered Species: The Coachella Valley Habitat Conservation Plan," in *The Ecological City: Preserving and Restoring Urban Biodiversity*, ed. Rutherford H. Platt, Rowan A. Rowntree, and Pamela C. Muick (Amherst: University of Massachusetts Press, 1994), 231–250; William Fulton, *The Reluctant Metropolis: The Politics of Urban Growth in Los Angeles* (Baltimore: Johns Hopkins University Press, 2001); Rutherford H. Platt, *Land Use and Society: Geography, Law, and Public Policy*, 3rd ed. (Washington, D.C.: Island Press, 2014); Jacob W. Malcom and Ya-Wei Li, "Data Contradict Common Perceptions about a Controversial Provision of the U.S. Endangered Species Act," *PNAS* 112 (2015): 15844–15849; George F. Wilhere, "The Role of Scientists in Statutory Interpretation of the U.S. Endangered Species Act," *CB* 31 (2017): 252–260; Christian Langpap, Joe Kerkvliet, and Jason F. Shogren, "The Economics of the U.S. Endangered Species Act: A Review of Recent Developments," *Review of Environmental Economics and Policy* 12 (2018): 69–91. Note that under California's state Endangered Species Act, the California gnatcatcher is listed as a "Species of Special Concern," which provides no legal protection; see Endangered Species Comm. of Bldg. Indus. Ass'n of S. Cal. v. Babbitt, 852 F.Supp. 32, 34 (D.D.C. 1994) at 36: "Prior to listing of the *P.c.c.*, many jurisdictions and landowners with control over *P.c.c.* habitat elected not to enroll in the program.... According to Director Beattie, absent the legal protection afforded through the listing of the *P.c.c.*, habitat loss and fragmentation can continue to occur prior to development and implementation of adequate conservation plans under the N.C.C.P.... The listing of the *P.c.c.* provides incentives for the voluntary enrollment of landowners in the N.C.C.P. program."

2. Dale D. Goble, J. Michael Scott, and Frank W. Davis, eds., *The Endangered Species Act at Thirty*, 2 vols. (Washington, D.C.: Island Press, 2005); Joel Greenberg, *A Feathered River across the Sky: The Passenger Pigeon's Flight to Extinction* (New York: Bloomsbury USA, 2014); Alvin Powell, *The Race to Save the World's Rarest Bird: The Discovery and Death of the Po'ouli* (Mechanicsburg, Pa.: Stackpole Books, 2008), 205–206; Greenberg, *Feathered River*.

3. J. Peyton Doub, *The Endangered Species Act: History, Implementation, Successes, and Controversies* (Boca Raton, Fla.: CRC Press, 2013); Douglas Brinkley, *The Wilderness Warrior: Theodore Roosevelt and the Crusade for America* (New York: Harper Perennial, 2010); Daniel M. Evans et al., "Species Recovery in the United States: Increasing the Effectiveness of the Endangered Species Act," *Issues in Ecology* 20 (2016): 1–28.

4. SELS, *The Endangered Species Act* (Stanford, Calif.: Stanford University Press, 2001); Michael J. Bean, "Historical Background of the Endangered Species Act," in *Endangered Species Act: Law, Policy, and Perspectives,* ed. Donald C. Baur and William Robert Irvin (Chicago: American Bar Association, 2010), 8–15; Doub, *Endangered Species Act;* Platt, *Land Use.*

5. Goble, Scott, and Davis, *Endangered Species Act at Thirty;* Shonil A. Bhagwat and Claudia Rutte, "Sacred Groves: Potential for Biodiversity Management," *Frontiers in Ecology and the Environment* 4 (2006): 519–524; Joe Roman, *Listed: Dispatches from America's Endangered Species Act* (Cambridge, Mass.: Harvard University Press, 2011); Doub, *Endangered Species Act;* David Vogel, *California Greenin': How the Golden State Became an Environmental Leader* (Princeton, N.J.: Princeton University Press, 2018); Evans et al., "Species Recovery"; SELS, *Endangered Species Act.*

6. 7 U.S.C. § 136; 16 U.S.C. § 1531 et seq.; with major amendments in 1978, 1982, 1988, and 2004. "Prioritized over economic considerations" and similar sentiments (e.g., "whatever the cost") have been weakened by congressional amendments over time. Beatley, "Reconciling Urban Growth"; Powell, *Race to Save;* Mark W. Schwartz, "The Performance of the Endangered Species Act," *Annual Review of Ecology, Evolution and Systematics* 39 (2008): 279–299; Daniel M. Evans, Dale D. Goble, and J. Michael Scott, "New Priorities as the Endangered Species Act Turns 40," *Frontiers in Ecology and the Environment* 11 (2013): 519; Goble, Scott, and Davis, *Endangered Species Act at Thirty;* Platt, *Land Use;* SELS, *Endangered Species Act;* Doub, *Endangered Species Act;* Langpap, Kerkvliet, and Shogren, "Economics"; John Buse, "Can a Multi-Species Habitat Conservation Plan Save San Diego's Vulnerable Vernal Pool Species?" *Golden Gate University Environmental Law Journal* 6 (2012): 53–80.

7. Jon Welner, "Natural Communities Conservation Planning: An Ecosystem Approach to Protecting Endangered Species," *Stanford Law Review* 47 (1995): 319–361; Timothy D. Male and Michael J. Bean, "Measuring Progress in U.S. Endangered Species Conservation," *Ecology Letters* 8 (2005): 986–992; J. Michael Scott et al., "By the Numbers," in Goble, Scott, and Davis, *Endangered Species Act at Thirty,* 1:16–35; Powell, *Race to Save;* Daniel M. Evans, Dale D. Goble, and J. Michael Scott, "New Priorities as the Endangered Species Act Turns 40," *Frontiers in Ecology and the Environment* 11 (2013): 519; Christopher Ketcham, "Inside the Effort to Kill Protections for Endangered Animals," *National Geographic.com,* May 19, 2017, http://news.natio nalgeographic.com/2017/05/endangered_speciesact/; Noah Greenwald et al., "Extinction and the U.S. Endangered Species Act," *PeerJ* 7 (2019): e6803, https://doi .org/10.7717/peerj.6803; Noah Greenwald, "Cherish the Act's Proven Power," *Nature* 504 (2014): 369; Ketcham, "Inside the Effort"; USFWS, *Environmental Conservation*

Online System, https://ecos.fws.gov/ecp0/reports/box-score-report, https://ecos.fws .gov/ecp0/reports/delisting-report; SELS, *Endangered Species Act.*

8. Evans, Goble, and Scott, "New Priorities"; Ketcham, "Inside the Effort"; USFWS, "Endangered and Threatened Wildlife and Plants; Removing the Bald Eagle in the Lower 48 States from the List of Endangered and Threatened Wildlife," 72 Fed. Reg. 37345–37372 (July 9, 2007); USFWS, "Endangered and Threatened Wildlife and Plants; Review of 2017 Final Rule, Greater Yellowstone Ecosystem Grizzly Bears," 83 Fed. Reg. 18737 (Apr. 30, 2018); USFWS, "Reclassification of American Alligator to Thr. Due to Similarity of Appearance throughout Remainder of Its Range," 52 Fed. Reg. 21059 (June 4, 1987); Roman, *Listed;* James W. Rivers et al., "An Analysis of Monthly Home Range Size in the Critically Endangered California Condor *Gymnogyps californianus*," *Bird Conservation International* 24 (2014): 492–504; Adrian P. Wydeven, Timothy R. van Deelen, and Edward Heske, eds., *Recovery of Gray Wolves in the Great Lakes Region of the United States: An Endangered Species Success Story* (New York: Springer, 2009); Kristina Alexander, *The Gray Wolf and the Endangered Species Act (ESA): A Brief Legal History* (Washington, D.C.: Congressional Research Service, July 27, 2011).

9. Lynn E. Dwyer, Dennis D. Murphy, and Paul R. Ehrlich, "Property Rights Case Law and the Challenge to the Endangered Species Act," *CB* 9 (1995): 725–741; Buse, "Multi-Species Habitat"; DeAnne Parker, "Comment: Natural Community Conservation Planning: California's Emerging Ecosystem Management Alternative," *University of Baltimore Journal of Environmental Law* 6 (1997): 107–140; David E. Moser, "Habitat Conservation Plans under the U.S. Endangered Species Act: The Legal Perspective," *Environmental Management* 26 (2000): S7–S13; Langpap, Kerkvliet, and Shogren, "Economics." Papers arguing against listing for the coastal California gnatcatcher often cite the economic costs of its listing incurred by landowners and the development industry; e.g., Robert M. Zink et al., "Genetics, Taxonomy, and Conservation of the Threatened California Gnatcatcher," *CB* 14 (2000): 1394–1405; John R. Skalski et al., "Type I Errors Linked to Faulty Statistical Analyses of Endangered Subspecies Classifications," *Journal of Agricultural, Biological, and Environmental Statistics* 13 (2008): 199–220; Zink et al., "Phylogeography"; Buse, "Multi-Species Habitat"; Stephanie Pincetl, Terry Watt, and Maria J. Santos, "Land Use Regulation for Resource Conservation," in *Ecosystems of California,* ed. Harold Mooney and Erika Zavaleta (Oakland: University of California Press, 2016), 899–924; Charles C. Mann and Mark L. Plummer, *Noah's Choice: The Future of Endangered Species* (New York: Alfred A. Knopf, 1995); Welner, "Natural Communities Conservation Planning"; Parker, "Comment"; SELS, *Endangered Species Act;* Doub, *Endangered Species Act.*

10. Fulton, *Reluctant Metropolis,* 208; Norris Scott, "Only 30: A Portrait of the Endangered Species Act as a Young Law," *Bioscience* 54 (2004): 288–294; Schwartz, "Performance," 293; Langpap, Kerkvliet, and Shogren, "Economics"; Michael L. Rosenzweig, *Win-Win Ecology: How the Earth's Species Can Survive in the Midst of Human Enterprise* (Oxford: Oxford University Press, 2003); Jonathan H. Alder, "The Leaky Ark: The Failure of Endangered Species Regulation on Private Land," in *Rebuilding the Ark: New Perspectives on Endangered Species Act Reform,* ed. Alder (Washington, D.C.: American Enterprise Institute Press, 2010), 6–31; Evans et al., "Species Recovery"; USFWS, "Determination of Threatened Status for the Coastal California Gnatcatcher," 58 Fed. Reg. 16742 (Mar. 30, 1993); Welner, "Natural Communities Conservation Planning"; John Tschirhart, "Account for Economics," *Nature* 504 (2013): 370; Langpap, Kerkvliet, and Shogren, "Economics"; Dwyer, Murphy, and Ehrlich, "Property Rights Case Law"; Thomas D. Feldman and Andrew E. G. Jonas, "Sage Scrub Revolution? Property Rights, Political Fragmentation, and Conservation Planning in Southern California under the Federal Endangered Species Act," *Annals of the Association of American Geographers* 90 (2000): 256–292; Eric Reitan, "Private Property Rights, Moral Extensionism and the Wise-Use Movement: A Rawlsian Analysis," *Environmental Values* 13 (2004): 329–347; Platt, *Land Use;* John D. Echeverria and Glenn P. Sugameli, "The Endangered Species Act and the Constitutional Takings Issue," in Baur and Irvin, *Endangered Species Act,* 292–315; Louis Sahagun, "Threatened Species Could Lose Habitat Protections under Department of the Interior Proposal," *LAT,* July 19, 2018.

11. Eric R. Glitzenstein, "Citizen Suits," in Baur and Irvin, *Endangered Species Act,* 260–291; Patrick Parenteau, "The Take Prohibition," in Baur and Irvin, *Endangered Species Act,* 146–159; Ketcham, "Inside the Effort"; "Politics of Extinction," *Center for Biological Diversity,* 2018, https://www.biologicaldiversity.org/campaigns/esa_attacks/trumptable.html; Brett Hartl, "Senate Bill Aims to Strip Protections from Nearly 1,100 Endangered Species," *Center for Biological Diversity,* Press Release, Sept. 28, 2017, https://www.biologicaldiversity.org/news/press_releases/2017/endangered-species-09-28-2017.php; S.1863 (115th Cong.); S.935 (115th Cong.) and H.R.3533 (113th Cong.); S.335 and H.R.3565 (115th Cong.); H.R.717 (115th Cong.).

12. Ketcham, "Inside the Effort."

13. Section numbers in this book always refer to the bill sections, not the statute. Readers interested in the details and definitions of the ESA can see three excellent books that explain them in legal terms, in their practical applications, and in their historical and broader societal impacts: SELS, *Endangered Species Act;* Doub, *Endangered Species Act;* and Baur and Irvin, *Endangered Species Act.*

14. J. B. Ruhl, "Listing Endangered and Threatened Species," in *Endangered Species Act: Law, Policy, and Perspectives,* ed. Donald C. Baur and Wm. Robert Irvin, 2nd ed. (Chicago: American Bar Association, 2010), 16–39. Lawsuits are filed in a court of law, whereas petitions are filed with the USFWS directly. Both usually involve trying to overturn a USFWS decision or to induce the USFWS to make one (e.g., a petition to add a species to the ESA list). Sometimes unsuccessful petitioners turn to the courts, but not always; normally one needs to exhaust the Services' petition process before progressing to a lawsuit (the petitioner's claim must be "ripe"; Echeverria and Sugameli, "Endangered Species Act and Constitutional Takings").

15. Holly Doremus, "Listing Decisions under the Endangered Species Act: Why Better Science Isn't Always Better Policy," *Washington University Law Review* 75 (1997): 1029–1153; Ruhl, "Listing Endangered and Threatened Species"; USFWS, "12-Month Finding on a Petition to Delist the Coastal California Gnatcatcher," 81 Fed. Reg. 59952 (Aug. 31, 2016). See Endangered Species Comm., 852 F.Supp. at 33–36; Ctr. for Envtl. Sci., Accuracy & Reliability v. U.S. Dep't of Interior, No. CV 17-2313 (JDB), 2019 WL 2870131 (D.D.C. July 3, 2019), at 1–2; Doremus, "Listing Decisions."

16. See Doremus, "Listing Decisions," on the advantages and disadvantages of "faith in science" to deliver objective policy decisions and for additional problems with the "best available science" standard. Sometimes the Services deviate from scientific consensus, usually without sufficient explanation for the inconsistency; Chevron U.S.A., Inc. v. Natural Res. Def. Council, Inc., 467 U.S. 837 (1984).

17. Ruhl, "Listing Endangered and Threatened Species"; Wyman, "Politics and Science"; Doub, *Endangered Species Act;* Evans et al., "Species Recovery."

18. USFWS, "Endangered and Threatened Wildlife and Plants; Special Rule Concerning Take of the Threatened Coastal California Gnatcatcher," 58 Fed. Reg. 65088 (Dec. 10, 1993). See Natural Res. Def. Council, Inc. v. U.S. Dep't of Interior, 275 F.Supp.2d 1136, 1138, 1156 (C.D. Cal. 2002); and Dwyer, Murphy, and Ehrlich, "Property Rights Case Law."

19. Ruhl, "Listing Endangered and Threatened Species," 21; USFWS, "Endangered and Threatened Wildlife and Plants; Regulations for Prohibitions to Threatened Wildlife and Plants," 84 Fed. Reg. 44753 (Aug. 27, 2019); USFWS and NMFS, "Endangered and Threatened Wildlife and Plants; Regulations for Listing Species and Designating Critical Habitat," 84 Fed. Reg. 45020 (Aug. 27, 2019); quote in Sahagun, "Threatened Species"; Lisa Friedman, Kendra Pierre-Louis, and Livia Albeck-Ripka, "Law That Saved the Bald Eagle Could Be Vastly Reworked," *New York Times,* July 19, 2018. House attempts included an amendment to the Department of Defense fund-

ing bill that would have stripped the endangered American burying beetle (*Nicrophorus americanus*) of ESA protections, similar to the snail darter: Jeremy Rehm, "U.S. Wildlife Law in Danger," *Nature* 560 (2018): 17–18; USFWS, "Endangered and Threatened Wildlife and Plants; Endangered Species Act Compensatory Mitigation Policy," 83 Fed. Reg. 36469 (July 30, 2018); Leia Larsen, "Federal Wildlife Director Steps Down from Post, Plans Return to Utah," *Standard-Examiner* (Ogden, Utah), Aug. 9, 2018; Julie Eilperin, Josh Dawsey, and Darryl Fears, "Interior Secretary Zinke Resigns amid Investigations," *Washington Post,* Dec. 15, 2018.

20. Endangered Species Act Amendments of 1978, Pub. L. No. 95-632, § 2, 92 Stat. 3751 (codified at 16 U.S.C. § 1532[5][A]); Schwartz, "Performance"; Federico Cheever, "Critical Habitat," in Baur and Irvin, *Endangered Species Act,* 40–69; SELS, *Endangered Species Act;* Amy Sinden, "The Economics of Endangered Species: Why Less Is More in the Economic Analysis of Critical Habitat Designations," *Harvard Environmental Law Review* 28 (2004): 129–214; Cheever, "Critical Habitat"; Langpap, Kerkvliet, and Shogren, "Economics." See, e.g., N.M. Cattle Growers Ass'n v. U.S. Fish & Wildlife Serv., 248 F.3d 1277 (10th Cir. 2001). Two cases involving the California gnatcatcher: Rancho Mission Viejo, L.L.C. v. Babbitt, CV 01-8412 SVW (CTx); and Bldg. Indus. Ass'n of S. Cal. v. Norton, CV 01-7028 SVW (CTx).

21. Norris, "Only 30," 291; Doub, *Endangered Species Act.*

22. U.S. Department of the Interior, Office of the Secretary, *Endangered Species Act "Broken"—Flood of Litigation over Critical Habitat Hinders Species Conservation,* News Brief, May 28, 2003; Katherine E. Gibbs and David J. Currie, "Protecting Endangered Species: Do the Main Legislative Tools Work?" *PLoS ONE* 7 (2012): https://doi.org/10.1371/journal.pone.0035730; Cheever, "Critical Habitat." Designations forced by lawsuits include: Conservation Council for Haw. v. Babbitt, 2 F.Supp.2d 1280 (D. Haw. 1998); Sierra Club v. U.S. Fish & Wildlife Serv., 911 F.3d 967 (9th Cir. 2018); Gifford Pinchot Task Force v. U.S. Fish & Wildlife Serv., 378 F.3d 1059 (9th Cir. Aug. 6, 2004), amended, 387 F.3d 968 (Oct. 28, 2004). Schwartz, "Performance"; USFWS, "Endangered and Threatened Wildlife and Plants; Determination of Whether Designation of Critical Habitat for the Coastal California Gnatcatcher Is Prudent," 64 Fed. Reg. 5957 (Feb. 8, 1999); Jared B. Fish, "Note: Critical Habitat Designations after New Mexico Cattle Growers: An Analysis of Agency Discretion to Exclude Critical Habitat," *Fordham Environmental Law Review* 21 (2010): 575–635; U.S. Department of the Interior, *"Broken";* Norris, "Only 30," 291; Sinden, "Economics of Endangered Species"; Martin F. J. Taylor, Kierán F. Suckling, and Jeffrey J. Rachlinski, "The Effectiveness of the Endangered Species Act: A Quantitative Analysis," *BioScience* 55 (2005): 360–367.

23. 64 Fed. Reg. 5757, 5958. This decision is also used when a species is threatened by overhunting or collecting and when publicizing the location of remaining individuals would exacerbate this risk. The not-prudent reason for refusing to designate critical habitat was quite common in the 1980s and 1990s, until a wave of lawsuits (including one over the coastal California gnatcatcher) changed this practice (NRDC, 275 F. Supp. 2d at 1141 (quoting Sierra Club v. U.S. Fish & Wildlife Serv., 245 F.3d 434, 437 (5th Cir. 2001) (footnote omitted)); Cheever, "Critical Habitat"; 58 Fed. Reg. 16742; Marla Cone, "Gnatcatcher Habitat Loss Called 'Significant,'" *LAT,* Apr. 21, 1992; Jonathan L. Atwood and David R. Bontrager, "California Gnatcatcher (*Polioptila californica*)," version 2.0, in *The Birds of North America,* ed. A. F. Poole and F. B. Gill (Cornell Lab of Ornithology, Ithaca, NY), 2001, https://doi.org/10.2173 /bna.574. See Natural Res. Def. Council, Inc. v. U.S. Dep't of Interior, 113 F.3d 1121, 1123, 1125 (9th Cir. 1997); USFWS, "Endangered and Threatened Wildlife and Plants; Final Determination of Critical Habitat for the Coastal California Gnatcatcher; Final Rule," 65 Fed. Reg. 63680 (Oct. 24, 2000). See NRDC, 275 F. Supp. 2d 1136, which argued that the critical habitat designation was too small (because it excluded NCCP, MSHCP, and HCP areas). Complaints filed in Rancho Mission Viejo, L.L.C. v. Babbitt, CV 01-8412 SVW (CTx) and Bldg. Ind. Ass'n of S. Cal. v. Norton, CV 01-7028 SVW (CTx) both argued that the critical habitat designation was too large given economic impacts. USFWS, "Endangered and Threatened Wildlife and Plants; Designation of Critical Habitat for the Coastal California Gnatcatcher (*Polioptila californica californica*) and Determination of Distinct Vertebrate Population Segment for the California Gnatcatcher (*Polioptila californica*); Proposed Rule," 68 Fed. Reg. 20228 (Apr. 24, 2003).

24. U.S. Department of the Interior, *"Broken";* Cheever, "Critical Habitat"; U.S. General Accounting Office, "Information on How Funds Are Allocated and What Activities Are Emphasized," GAO-02-581 (2002); U.S. General Accounting Office, "Fish and Wildlife Service Uses Best Available Science to Make Listing Decisions, but Additional Guidance Needed for Critical Habitat Designations," GAO-03-803 (2003).

25. 72 Fed. Reg. 72010, 72021–72022: $459,907,538 Draft Economic Analysis estimate divided by 2003 critical habitat area of 71,516 hectares (table 1); Daniel S. Cooper, Jennifer Mongolo, and Chris Dellith, "Status of the California Gnatcatcher at the Northern End of Its Range," *WB* 48 (2017): 124–140.

26. 64 Fed. Reg. 5957, 5958; Cheever, "Critical Habitat"; 16 U.S.C. § 1532(19); 50 C.F.R. §§ 17.31(a), 17.21(c), 17.3; SELS, *Endangered Species Act.* See NRDC, 275 F. Supp. 2d at 1145 n.18: "Unlike the Section 9 take prohibition, the critical habi-

tat designation only imposes Section 7 federal agency consultation for actions with a federal nexus." National Association of Home Builders, *Developer's Guide to Endangered Species Regulation* (Washington, D.C.: Home Builders Press, 1996), 109; Roman, *Listed*.

27. Jeffrey J. Rachlinski, "Noah by the Numbers: An Empirical Evaluation of the Endangered Species Act," *Cornell Law Review* 82 (1997): 356–389; Male and Bean, "Measuring Progress"; Martin F. J. Taylor, Kierán F. Suckling, and Jeffrey J. Rachlinski, "The Effectiveness of the Endangered Species Act: A Quantitative Analysis," *BioScience* 55 (2005): 360–367; Kierán F. Suckling and Martin Taylor, "Critical Habitat and Recovery," in Goble, Scott, and Davis, *Endangered Species Act at Thirty*, 1:75–89; Joe Kerkvliet and Christian Langpap, "Learning from Endangered and Threatened Species Recovery Programs: A Case Study Using U.S. Endangered Species Act Recovery Scores," *Ecological Economics* 63 (2007): 499–510; Gibbs and Currie, "Protecting Endangered Species"; Abbey E. Camaclang et al., "Current Practices in the Identification of Critical Habitat for Threatened Species," *CB* 29 (2015): 482–492; Jeffrey E. Zabel and Robert W. Paterson, "The Effects of Critical Habitat Designation on Housing Supply: An Analysis of California Housing Construction Activity," *Journal of Regulatory Science* 46 (2006): 67–95; John M. Quigley and Aaron M. Swoboda, "The Urban Impacts of the Endangered Species Act: A General Equilibrium Analysis," *Journal of Urban Economics* 61 (2007): 299–318; Erik J. Nelson et al., "Identifying the Impacts of Critical Habitat Designation on Land Cover Change," *Resource and Energy Economics* 47 (2017): 89–125; Sahagun, "Threatened Species"; Economic & Planning Systems, *Economic Analysis of Critical Habitat Designation for the California Gnatcatcher* (Division of Economics, USFWS, Feb. 24, 2004); Economic & Planning Systems, *Report Addendum: Economic Analysis of Critical Habitat Designation for the California Gnatcatcher* (Division of Economics, USFWS, Sept. 14, 2007). See NRDC, 275 F. Supp. 2d at 1138–40.

28. Dale D. Goble, "Recovery," in Baur and Irvin, *Endangered Species Act,* 70–103; Wilhere, "Role of Scientists"; SELS, *Endangered Species Act*.

29. National Research Council, Committee on Scientific Issues in the Endangered Species Act, *Science and the Endangered Species Act* (Washington, D.C.: National Academies Press, 1995); Timothy H. Tear et al., "Status and Prospects for Success of the Endangered Species Act: A Look at Recovery Plans," *Science* 262 (1993): 976–977; Timothy H. Tear et al., "Recovery Plans and the Endangered Species Act: Are Criticisms Supported by Data?" *CB* 9 (1995): 182–195; Goble, "Recovery"; Evans et al., "Species Recovery"; Taylor et al., "Effectiveness"; Kerkvliet and Langpap, "Learning." Species covered in multispecies recovery plans are more likely to be in decline, how-

ever: Taylor et al., "Effectiveness"; P. Dee Boersma et al., "How Good Are Endangered Species Recovery Plans?" *BioScience* 51 (2001): 643–649; J. Alan Clark et al., "Improving U.S. Endangered Species Act Recovery Plans: Key Findings and Recommendations of the SCB Recovery Plan Project," *CB* 16 (2002): 1510–1519; William E. Kunin and Kevin J. Gaston, eds., *The Biology of Rarity: Causes and Consequences of Rare-Common Differences* (Dordrecht, Netherlands: Springer, 1997).

30. Marco Restani and John M. Marzluff, "Funding Extinction? Biological Needs and Political Realities in the Allocation of Resources to Endangered Species Recovery," *BioScience* 52 (2002): 169–177; David L. Leonard Jr., "Recovery Expenditures for Birds Listed under the US Endangered Species Act: The Disparity between Mainland and Hawaiian Taxa," *BC* 141 (2008): 2054–2061; Julie K. Miller et al., "The Endangered Species Act: Dollars and Sense?" *BioScience* 52 (2002): 163–168; Norris, "Only 30"; Powell, *Race to Save;* Schwartz, "Performance"; J. R. DeShazo and Jody Freeman, "Congressional Politics," in Goble, Scott, and Davis, *Endangered Species Act at Thirty,* 1:68–73.

31. Norris, "Only 30"; Schwartz, "Performance"; Michael L. Morrison, "Facilitating Development of Multiple-Species Conservation Reserves and Habitat Conservation Plans: A Synthesis of Recommendations," *Environmental Management* 26 (2000): S3–S6; Evans et al., "Species Recovery"; SELS, *Endangered Species Act;* Robert W. McFarlane, *Stillness in the Pines: The Ecology of the Red Cockaded Woodpecker* (New York: W. W. Norton, 1994). See Sierra Club v. Yeutter, 926 F.2d 429 (5th Cir. 1991).

32. USFWS, *Report to Congress: Recovery Program, Endangered and Threatened Species: 1994,* https://www.fws.gov/endangered/esa-library/pdf/1994_USFWS_Recover_Reports.pdf; USFWS, *Report to Congress on the Recovery of Threatened and Endangered Species: Fiscal Years 2013–2014,* https://www.fws.gov/endangered/esa-library/pdf/Recovery_Report_FY2013-2014.pdf; USFWS, "Endangered and Threatened Species Listing and Recovery Priority Guidelines," 48 Fed. Reg. 43104 (Sept. 21, 1983).

33. Norris, "Only 30"; Thornton, "Searching for Consensus"; Parker, "Comment"; Moser, "Habitat Conservation Plans"; SELS, *Endangered Species Act;* Patrick W. Ryan and Erika E. Malmen, "Interagency Consultation under Section 7," in Baur and Irvin, *Endangered Species Act,* 104–125. The 2004 National Defense Authorization Act grants the Department of Defense an exemption from ESA critical habitat designations on military property, as long as the area is managed under an Integrated Natural Resources Management Plan; William E. Sitzabee, Charles A. Bleckmann, and Ellen C. England, "An Evaluation of Endangered Species Act Exemptions in the Department of Defense and the U.S. Air Force," *Federal Facilities Environmental Journal* 15 (2004):

19–28; Doub, *Endangered Species Act;* Zygmunt J. B. Plater, *The Snail Darter and the Dam: How Pork-Barrel Politics Endangered a Little Fish and Killed a River* (New Haven: Yale University Press, 2014); Platt, *Land Use;* Ruhl, "Listing Endangered and Threatened Species"; Atwood and Noss, "Gnatcatchers and Development"; Malcom and Li, "Data Contradict."

34. 81 Fed. Reg. 59952. See Laguna Greenbeld, Inc. v. U.S. Dep't of Transp., 42 F.3d 517 (9th Cir. 1994), as amended on denial of reh'g (Dec. 20, 1994).

35. 81 Fed. Reg. 59952; U.S. Marine Corps, *Final Joint Integrated Natural Resources Management Plan for Marine Corps Base and Marine Corps Air Station Camp Pendleton, California* (San Diego, 2018), https://www.pendleton.marines.mil/Portals/98 /Docs/Environmental/Natural%20Resources/Integrated_Natural_Resources_Man agement_Plan_2018.pdf; Bruce A. Stein, Cameron Scott, Nancy Benton, "Federal Lands and Endangered Species: The Role of Military and Other Federal Lands in Sustaining Biodiversity," *BioScience* 58 (2008): 339–347.

36. Norris, "Only 30"; Mann and Plummer, *Noah's Choice;* Dwyer, Murphy, and Ehrlich, "Property Rights Case Law"; Tony Davis, "Critical Habitat: The Inside Story," *High Country News,* Feb. 20, 2006; Ryan and Malmen, "Interagency Consultation"; Plater, *Snail Darter.* More accurately, Gingrich voted against the dam, a notable pork barrel boondoggle: Cong. Rec., House, June 16, 1978, 17936; Roman, *Listed;* Larry Bowers, "The 1970s: Days of the Dam and the Darter," *Cleveland Daily Banner,* Oct. 21, 2015.

37. Thornton, "Searching for Consensus"; Parker, "Comment"; SELS, *Endangered Species Act;* USFWS, Endangered Species Program, "Habitat Conservation Plans under the Endangered Species Act" (2011), https://www.fws.gov/endangered /esa-library/pdf/hcp.pdf.

38. Dwyer, Murphy, and Ehrlich, "Property Rights Case Law"; SELS, *Endangered Species Act;* Parenteau, "Take Prohibition"; see Palila v. Haw. Dep't of Land & Natural Res., 852 F.2d. 1106 (9th Cir. 1988); Marbled Murrelet v. Babbitt, 83 F.3d. 1060 (9th Cir. May 7, 1996), as amended on denial of reh'g (June 26, 1996); A. Dan Tarlock, "Fred Bosselman as Participant-Observer Lawyer: The Case of Habitat Conservation Planning," *Florida State University Journal of Land Use and Environmental Law* 17 (2018): 43–55. See Babbitt v. Sweet Home Chapter of Communities for a Great Or., 515 U.S. 687, 707–709 (1995); N.M. Cattle Growers Ass'n, 248 F.3d 1277; Thornton, "Searching for Consensus"; Moser, "Habitat Conservation Plans." See Strahan v. Coxe, 127 F.3d. 155 (1st Cir. 1997); U.S. v. Town of Plymouth, Mass., 6 F.Supp.2d 81 (D. Mass. 1998); Loggerhead Turtle v. Cnty. Council of Volusia Cnty., Fla., 896 F.Supp. 1170 (M.D. Fla. 1995).

39. Thornton, "Searching for Consensus"; David W. Myers, "Ire after the Fire: Victims Say Endangered Rat Got More Protection Than Their Homes," *LAT,* Nov. 16, 1993; Mike Davis, *Ecology of Fear: Los Angeles and the Imagination of Disaster* (New York: Vintage Books, 1998), 134.

40. Moser, "Habitat Conservation Plans"; Thornton, "Searching for Consensus"; Dwyer, Murphy, and Ehrlich, "Property Rights Case Law"; SELS, *Endangered Species Act;* Craig W. Thomas, *Bureaucratic Landscapes: Interagency Cooperation and the Preservation of Biodiversity* (Cambridge, Mass.: MIT Press, 2003); USFWS and NMFS, *Habitat Conservation Planning and Incidental Take Permit Processing Handbook* (2016), https://www.fws.gov/endangered/esa-library/pdf/HCP_Handbook.pdf; Tarlock, "Fred Bosselman"; USFWS, Endangered Species Program, "Habitat Conservation Plans"; 81 Fed. Reg. 59952.

41. USFWS, "Habitat Conservation Plan Assurances ('No Surprises') Rule," 63 Fed. Reg. 8859 (Feb. 23, 1998); USFWS, "Safe Harbor Agreements and Candidate Conservation Agreements with Assurances," 64 Fed. Reg. 32706 (June 17, 1999); Dwyer, Murphy, and Ehrlich, "Property Rights Case Law"; SELS, *Endangered Species Act;* Rosenzweig, *Win-Win Ecology;* Norris, "Only 30"; Doub, *Endangered Species Act;* Evans et al., "Species Recovery"; Tarlock, "Fred Bosselman"; Jonathan H. Adler, "The Adverse Environmental Consequences of Uncompensated Land-Use Controls," in *Property Rights: Eminent Domain and Regulatory Takings Re-Examined,* ed. Bruce L. Benson (New York: Independent Institute, 2010), 187–210; Alejandro E. Camacho, Elizabeth Taylor, and Melissa Kelly, "Lessons from Area-Wide, Multi-Agency Habitat Conservation Plans in California," *ELR* 46 (2016): 10222–10248; Fred P. Bosselman, "The Statutory and Constitutional Mandate for a No Surprises Policy," *Ecology Law Quarterly* 24 (1997): 717. But see Daniel A. Farber, "Taking Slippage Seriously: Noncompliance and Creative Compliance in Environmental Law," *Harvard Environmental Law Review* 23 (1999): 297–325.

42. SELS, *Endangered Species Act;* David S. Wilcove et al., *Rebuilding the Ark: Towards a More Effective Endangered Species Act for Private Land* (New York: Environmental Defense Fund, 1996); Rosenzweig, *Win Win Ecology;* USFWS, Endangered Species Program, "Habitat Conservation Plans." But see Melinda Taylor and Holly Doremus, "Habitat Conservation Plans and Climate Change: Recommendations for Policy," *ELR* 45 (2015): 10866: requiring additional mitigation from a permittee or the USFWS depends on whether there have been "changed circumstances" (for which permittees can be required to do additional mitigation), "unforeseen circumstances" (for which they cannot), or "extraordinary circumstances," which allows the USFWS to negotiate different or additional management actions on enrolled hectares. The cir-

cumstances depend on the context and USFWS discretion. Daniel Pollak, *The Future of Habitat Conservation? The NCCP Experience in Southern California* (Sacramento: California Research Bureau CRB-01-009, 2001); Parker, "Comment"; Moser, "Habitat Conservation Plans"; Douglas P. Wheeler and Ryan M. Rowberry, "Habitat Conservation Plans and the Endangered Species Act," in Baur and Irvin, *Endangered Species Act,* 220–245.

43. See Spirit of Sage Council v. Norton, 294 F. Supp. 2d 67, 72 (D.D.C. 2003), *amended sub nom.,* Spirit of Sage Council v. Norton, No. CIV. A. 98–1873, 2004 WL 1326279 (D.D.C. June 10, 2004), *and appeal dismissed, judgment vacated in part sub nom.,* Spirit of Sage Council v. Norton, 411 F.3d 225 (D.C. Cir. 2005); USFWS, "Endangered and Threatened Species: Incidental Take Permit Revocation Regulations," 69 Fed. Reg. 71723 (Dec. 10, 2004).

44. Mission blue butterfly (*Icaricia icarioides missionensis*) and San Bruno elfin butterfly (*Callophrys mossii bayensis*); Thornton, "Searching for Consensus"; Welner, "Natural Communities Conservation Planning"; Norris, "Only 30"; Langpap, Kerkvliet, and Shogren, "Economics"; Fred Bosselman, "Planning to Prevent Species Endangerment," *Land Use Law and Zoning Digest* 44 (1992): 3–8; Moser, "Habitat Conservation Plans"; Daniel Pollak, *Natural Community Conservation Planning (NCCP): The Origins of an Ambitious Experiment to Protect Ecosystems* (Sacramento: California Research Bureau CRB-01-002, 2001); Thomas, *Bureaucratic Landscapes.*

45. In California, the HCP parties include the USFWS and California Department of Fish and Wildlife, plus the landowners and other stakeholders (such as environmental and conservation groups). Thornton, "Searching for Consensus"; Langpap, Kerkvliet, and Shogren, "Economics"; Thomas, *Bureaucratic Landscapes;* Schwartz, "Performance"; Moser, "Habitat Conservation Plans."

46. Thornton, "Searching for Consensus"; Thomas, *Bureaucratic Landscapes;* Matthew E. Rahn, Holly Doremus, and James Diffendorfer, "Species Coverage in Multispecies Habitat Conservation Plans: Where's the Science?" *BioScience* 56 (2006): 613–619; Clark S. Winchell and Paul F. Doherty Jr., "Using California Gnatcatcher to Test Underlying Models in Habitat Conservation Plans," *Journal of Wildlife Management* 72 (2008): 1322–1327; Schwartz, "Performance."

47. Columbia University Integrative Capstone Workshop in Sustainability Management, "Understanding the Species Mitigation Market in the United States," prepared for the Nature Conservancy, faculty advisor: George Sarrinikolaou (New York: Columbia University, 2016); Christian Langpap and Joe Kerkvliet, "Endangered Species Conservation on Private Land: Assessing the Effectiveness of Habitat Conservation Plans," *Journal of Environmental Economics and Management* 64 (2012):

1–15; County of San Mateo Parks Department, *San Bruno Mountain Park Master Plan Draft 2001*, https://parks.smcgov.org/sites/parks.smcgov.org/files/documents /files/San%20Bruno%20Mountain%20Master%20Plan_0.pdf. Recreation may not be compatible with all species; negative impacts of recreation activities on orange-throated whiptails and other conservation target species have been noted in many NCCP/HCP preserves: Elizabeth Lucas, "Recreation-Related Disturbance to Wildlife in California: Better Planning for and Management of Recreation Are Vital to Conserve Wildlife in Protected Areas Where Recreation Occurs," *California Fish and Wildlife Journal, Recreation Special Issue* (2020): 29–51.

48. Thornton, "Searching for Consensus"; Welner, "Natural Communities Conservation Planning"; Evans, Goble, and Scott, "New Priorities"; Mann and Plummer, *Noah's Choice;* Richard W. Pombo, *The ESA at 30: Time for Congress to Update and Strengthen the Law*, Committee Report, U.S. House of Representatives Committee on Resources (2004); Schwartz, "Performance"; Gibbs and Currie, "Protecting Endangered Species."

49. Rachlinski, "Noah by the Numbers"; Evans et al., "Species Recovery"; Langpap, Kerkvliet, and Shogren, "Economics"; Schwartz, "Performance"; Roman, *Listed;* Noel Greenwald et al., *Shortchanged: Funding Needed to Save America's Most Endangered Species,* December 2016, https://www.biologicaldiversity.org/programs/biodi versity/pdfs/Shortchanged.pdf; Abel Valdivia, Shaye Wolf, and Kierán Suckling, "Marine Mammals and Sea Turtles Listed under the U.S. Endangered Species Act Are Recovering," *PLoS ONE* 14 (2019): https://doi.org/10.1371/journal.pone.0210164; Greenwald et al., "Extinction"; Male and Bean, "Measuring Progress"; Taylor et al., "Effectiveness"; Gibbs and Currie, "Protecting Endangered Species"; Amy W. Ando, "Focus on the Bigger Picture," *Nature* 504 (2014): 369; Ketcham, "Inside the Effort."

50. Kierán Suckling et al., *A Wild Success: A Systematic Review of Bird Recovery under the Endangered Species Act* (Center for Biological Diversity, June 2016).

51. Taylor et al., "Effectiveness"; Schwartz, "Performance"; Gibbs and Currie, "Protecting Endangered Species"; SELS, *Endangered Species Act;* Evans et al., "Species Recovery"; Ketcham, "Inside the Effort"; Welner, "Natural Communities Conservation Planning"; Greenwald et al., *Shortchanged.*

52. Evans, Goble, and Scott, "New Priorities"; Platt, *Land Use;* Welner, "Natural Communities Conservation Planning"; Ando, "Focus"; Evans, Goble, and Scott, "New Priorities"; Evans et al., "Species Recovery"; Holly Doremus and Joel E. Pagel, "Why Listing May Be Forever: Perspectives on Delisting under the U.S. Endangered Species Act," *CB* 15 (2001): 1258–1268; Schwartz, "Performance."

53. Dwyer, Murphy, and Ehrlich, "Property Rights Case Law"; Norris, "Only

30"; Welner, "Natural Communities Conservation Planning"; Carsten Mann and James D. Absher, "Adjusting Policy to Institutional, Cultural and Biophysical Context Conditions: The Case of Conservation Banking in California," *Land Use Policy* 36 (2014): 73–82; Rui Santos et al., "Reviewing the Role of Habitat Banking and Tradable Development Rights in the Conservation Policy Mix," *Environmental Conservation* 42 (2015): 294–305; Langpap, Kerkvliet, and Shogren, "Economics"; James Boyd and Rebecca S. Epanchin-Neill, *Private Sector Conservation Investments under the Endangered Species Act* (Washington, D.C.: Resources for the Future, RFF-DP-17-11, 2017).

Chapter 6. NCCP to the Rescue

1. Maria João Santos, Terry Watt, and Stephanie Pincetl, "The Push and Pull of Land Use Policy: Reconstructing 150 Years of Development and Conservation Land Acquisition," *PLoS ONE* 9 (2014): https://doi.org/10.1371/journal.pone.0103489; Stephanie Pincetl, Terry Watt, and Maria J. Santos, "Land Use Regulation for Resource Conservation," in *Ecosystems of California,* ed. Harold Mooney and Erika Zavaleta (Oakland: University of California Press, 2016), 899–924.

2. Tom Kenworthy, "Babbitt Clears Compromise to Protect California Bird," *Washington Post,* May 26, 1993; Jeanne Clark, "NCCP: A New Approach to Saving Species," *Outdoor California,* March–April 1994, 6–7; Tony Davis, "High Noon for Habitat," *High Country News,* Feb. 20, 2006; Pincetl, Watt, and Santos, "Land Use Regulation"; A. Dan Tarlock, "Fred Bosselman as Participant-Observer Lawyer: The Case of Habitat Conservation Planning," *Florida State University Journal of Land Use and Environmental Law* 17 (2018): 43–55.

3. Jonathan L. Atwood and Reed F. Noss, "Gnatcatchers and Development: A 'Train Wreck' Avoided?" *Illahee* 10 (1994): 123–130; Thomas S. Reid and Dennis D. Murphy, "Providing a Regional Context for Local Conservation Action," *BioScience* 45 (1995): S84–S90; Jon Welner, "Natural Communities Conservation Planning: An Ecosystem Approach to Protecting Endangered Species," *Stanford Law Review* 47 (1995): 319–361; David W. Stevens, "Utility Participation in a Multispecies Plan," *Environmental Management* 20 (1996): 841–847; Marc J. Ebbin, "Is the Southern California Approach to Conservation Succeeding?" *Ecology Law Quarterly* 24 (1997): 695–706; DeAnne Parker, "Comment: Natural Community Conservation Planning: California's Emerging Ecosystem Management Alternative," *University of Baltimore Journal of Environmental Law* 6 (1997): 107–140; Santos, Watt, and Pincetl, "Push and Pull"; Daniel Pollak, *The Future of Habitat Conservation? The NCCP Experience in Southern California* (Sacramento: California Research Bureau CRB-01-009, 2001);

Welner, "Natural Communities Conservation Planning"; USFWS, "12-Month Finding on a Petition to Delist the Coastal California Gnatcatcher," 81 Fed. Reg. 59952 (Aug. 31, 2016); Atwood and Noss, "Gnatcatchers and Development"; Michael A. Mantell, "Beyond Single Species: The California Experiment," *Illahee* 10 (1994): 131–135; Welner, "Natural Communities Conservation Planning"; Stevens, "Utility Participation"; Douglas J. Wheeler, "The Ecosystem Approach: New Departures for Land and Water: Keynote Address," *Ecology Law Quarterly* 24 (1997): 623–630; Stephanie Pincetl, "The Preservation of Nature at the Urban Fringe," in *Up against the Sprawl: Public Policy and the Making of Southern California,* ed. Jennifer Wolch, Manuel Pastor Jr., and Peter Dreier (Minneapolis: University of Minnesota Press, 2004), 225–251; Clark, "NCCP," 6–9, emphasis in the original.

4. Robert D. Thornton, "Searching for Consensus and Predictability: Habitat Conservation Planning under the Endangered Species Act of 1973," *Environmental Law* 21 (1991): 605–656; William Fulton, *The Reluctant Metropolis: The Politics of Urban Growth in Los Angeles* (Baltimore: Johns Hopkins University Press, 2001), 115–117, 212. The Metropolitan Water District (MWD) has been a central figure in both urban sprawl (since water supply is necessary for growth) and conservation efforts. The MWD contributed $15 million plus all the land surrounding a planned reservoir (in Domenigoni valley, now called the "Eastside Reservoir") for Stephens' kangaroo rat habitat, allowing some housing development (and habitat destruction) to proceed in Riverside County. The land surrounding two of its other reservoirs (Lake Mathews and Lake Skinner) are preserved for the California gnatcatcher. Welner, "Natural Communities Conservation Planning"; Ebbin, "Southern California Approach"; Fred Bosselman, "Planning to Prevent Species Endangerment," *Land Use Law and Zoning Digest* 44 (1992): 3–8; Adrianna Kripke, "Conservation Planning in Orange County, California: Linking Ecosystem Protection to Open Space Preservation," *ELR* 36 (2006): 10073; Daniel Pollak, *Natural Community Conservation Planning (NCCP): The Origins of an Ambitious Experiment to Protect Ecosystems* (Sacramento: California Research Bureau CRB-01-002, 2001); Pollak, *The Future of Habitat Conservation?,* Pincetl, "Urban Fringe."

5. Atwood and Noss, "Gnatcatchers and Development," 129; Jonathan L. Atwood, "Gnatcatchers and the NCCP: Will We Know When the Species Should Be Delisted?" in *Second Interface between Ecology and Land Development in California,* ed. Jon E. Keeley, Melanie Baer-Keeley, and C. J. Fotheringham, Open-File Report 00-62 (Sacramento: U.S. Geological Survey, Western Ecological Research Center, 2000), 179.

6. California Fish and Game Code, Division 3, Chapter 10. Natural Commu-

nity Conservation Planning Act [2800–2835]; Sarah E. Reed et al., *Wildlife Response to Human Recreation on NCCP Reserves in San Diego County,* Wildlife Conservation Society, Final Report, Agreement No. P1182112 (https://nrm.dfg.ca.gov/FileHandler .ashx?DocumentID=99222, 2014); Daniel M. Evans et al., "Species Recovery in the United States: Increasing the Effectiveness of the Endangered Species Act," *Issues in Ecology* 20 (2016): 1–28; Douglas P. Wheeler, "Charting a Course in the Golden State," in *Past Meets Future: Saving America's Historic Environments,* ed. Antoinette J. Lee (Washington, D.C.: National Trust for Historic Preservation and Preservation Press, 1992); Thomas D. Feldman and Andrew E. G. Jonas, "Sage Scrub Revolution? Property Rights, Political Fragmentation, and Conservation Planning in Southern California under the Federal Endangered Species Act," *Annals of the Association of American Geographers* 90 (2000): 256–292; USFWS, "Endangered and Threatened Wildlife and Plants; Special Rule Concerning Take of the Threatened Coastal California Gnatcatcher," 58 Fed. Reg. 65088 (Dec. 10, 1993); Atwood and Noss, "Gnatcatchers and Development"; Parker, "Comment"; Atwood, "Gnatcatchers"; Wheeler and Rowberry, "Habitat Conservation Plans"; Tarlock, "Fred Bosselman," 47–49; Pollak, *Future of Habitat Conservation?;* Fulton, *Reluctant Metropolis,* 202, 217; Pollak, *NCCP,* 17.

7. Stevens, "Utility Participation"; Pincetl, Watt, and Santos, "Land Use Regulation"; J. B. Ruhl, "Listing Endangered and Threatened Species," in *Endangered Species Act: Law, Policy, and Perspectives,* ed. Donald C. Baur and Wm. Robert Irvin, 2nd ed. (Chicago: American Bar Association, 2010), 16–39.

8. Christopher S. Galik and Lydia P. Olander, "Facilitating Markets and Mitigation: A Systematic Review of Early-Action Incentives in the U.S.," *Land Use Policy* 72 (2018): 1–11; Niak Sian Koh, Thomas Hahn, and Wiebren J. Boonstra, "How Much of a Market Is Involved in a Biodiversity Offset? A Typology of Biodiversity Offset Policies," *Journal of Environmental Management* 232 (2019): 679–691; California Resource Agency and California Environmental Protection Agency, "Official Policy on Conservation Banks," 1995, https://www.wildlife.ca.gov/Conservation/Planning/Ba nking/Laws-and-Policies; USFWS and California Department of Fish and Game, "Supplemental Policy Regarding Conservation Banks within the NCCP Area of Southern California," 1996, https://www.fws.gov/endangered/landowners/toolbox /Bank_Enabling_Instrument_Template-CA.doc (this explicitly linked conservation banking to the NCCP); USFWS, "Guidance for the Establishment, Use, and Operation of Conservation Banks," 68 Fed. Reg. 24753 (May 8, 2003) (this was based on California guidelines); H. Reşit Akçakaya and Jonathan L. Atwood, "A Habitat-Based Metapopulation Model of the California Gnatcatcher," *CB* 11 (1997): 422–434;

Rui Santos et al., "Reviewing the Role of Habitat Banking and Tradable Development Rights in the Conservation Policy Mix," *Environmental Conservation* 42 (2015): 294–305; Evans et al., "Species Recovery"; Carsten Mann and James D. Absher, "Adjusting Policy to Institutional, Cultural and Biophysical Context Conditions: The Case of Conservation Banking in California," *Land Use Policy* 36 (2014): 73–82; David A. Bunn, Peter B. Moyle, and Christine K. Johnson, "Maximizing the Ecological Contribution of Conservation Banks," *Wildlife Society Bulletin* 38 (2014): 377–385; Deborah Sullivan Brennan, "Wildlife Officials, Mountain Bikers Fight Turf War over Carlsbad Reserve," *San Diego Union-Tribune,* Apr. 13, 2019. See San Bernardino Valley Audubon Soc. v. Metro. Water Dist., 71 Cal.App.4th 382 (1999); Lydia P. Olander and Ben L. Young, "Integrating Large-Scale Planning into Environmental Markets and Related Programs: Status and Trends," NI WP 17-03 (Durham, N.C.: Duke University, Nicholas School of the Environment, 2016).

9. Pollak, *NCCP,* as per California Fish and Game Code § 2061; Kripke, "Conservation Planning"; Alejandro E. Camacho, Elizabeth Taylor, and Melissa Kelly, "Lessons from Area-Wide, Multi-Agency Habitat Conservation Plans in California," *ELR* 46 (2016): 10222–10248; Mantell, "Beyond Single Species"; Welner, "Natural Communities Conservation Planning"; Stevens, "Utility Participation." See Natural Res. Def. Council, Inc. v. U.S. Dep't of Interior, 113 F.3d 1121 (9th Cir. 1997); Pollak, *NCCP.*

10. Atwood and Noss, "Gnatcatchers and Development," 126; Fulton, *Reluctant Metropolis,* 213; Tarlock, "Fred Bosselman," 47; Audrey L. Mayer and Päivi M. Tikka, "Biodiversity Conservation Incentive Programs for Privately Owned Forests," *Environmental Science and Policy* 9 (2006): 614–625; Mark W. Schwartz, "The Performance of the Endangered Species Act," *Annual Review of Ecology, Evolution and Systematics* 39 (2008): 279–299. From *NRDC,* 113 F.3d 1121: "In any event, the NCCP alternative cannot be viewed as a functional substitute for critical habitat designation. Critical habitat designation triggers mandatory consultation requirements for federal agency actions involving critical habitat. The NCCP alternative, in contrast, is a purely voluntary program that applies only to non federal land-use activities. The Service itself recognized at the time of its final listing decision that 'no substantive protection of the coastal California gnatcatcher is currently provided by city/county enrollments [in the NCCP].' 58 Fed. Reg. at 16754. Accordingly, we reject the defendants' post hoc invocation of the NCCP to justify the Service's failure to designate critical habitat."

11. USFWS, "Determination of Endangered Status for Stephens' Kangaroo Rat," 53 Fed. Reg. 38465 (Sept. 30, 1988).

12. Jeanne Boyer, "Rat Race Gnaws at the Nerves of Riverside County Home Builders," *LAT,* June 18, 1989; Fulton, *Reluctant Metropolis,* 209–210; David Danelski, "Environment: Is Stephens' Kangaroo Rat Still Endangered?" *Press-Enterprise* (Riverside, CA), Sept. 3, 2015.

13. Thornton, "Searching for Consensus"; Kripke, "Conservation Planning"; Evans et al., "Species Recovery."

14. Pollak, *NCCP;* Kripke, "Conservation Planning."

15. William J. Clinton, "Remarks at a Town Meeting in San Diego, May 17, 1993," https://www.govinfo.gov/content/pkg/PPP-1993-book1/pdf/PPP-1993-book1-doc-pg680.pdf; Michael Flagg and John O'Dell, "Builders' Study Warns of Gnatcatcher Havoc," *LAT,* Aug. 29, 1991; Robert Reinhold, "Tiny Songbird Poses Big Test of U.S. Environmental Policy," *New York Times,* Mar. 16, 1993; Fulton, *Reluctant Metropolis,* 212; Jennifer Wolch, Manuel Pastor Jr., and Peter Dreier, "Introduction," in Wolch, Pastor, and Dreier, *Up against the Sprawl,* 12; Clinton, "Remarks," 691.

16. Cheryl Downey, "Toll-Road Agency Itching to Build on Burned Land," *Orange County Register,* Nov. 18, 1993; Fulton, *Reluctant Metropolis,* 202–204, 212, 218. See Laguna Greenbeld, Inc., 42 F.3d 517.

17. "Less public" means that the USFWS biological opinion was still unpublished and available only on request as of the writing of this book: USFWS, "Reinitiation of Formal Consultation on Implementation of the Special Rule for the Coastal California Gnatcatcher (1-6-93-FW-37RI)" (Unpublished report, prepared for USFWS, California State Supervisor, Sacramento, Oct. 18, 1996); Atwood, "Gnatcatchers"; 58 Fed. Reg. 65088; Pollak, *Future of Habitat Conservation?* See Spirit of Sage Council, 294 F.Supp.2d 67.

18. Atwood, "Gnatcatchers"; USFWS, "Reinitiation of Formal Consultation on Implementation of the Special Rule for the Coastal California Gnatcatcher (1-6-93-FW-37RI)" (Report prepared for USFWS, California State Supervisor, Sacramento, 1996), 6 (both quotations).

19. J. B. Kirkpatrick and C. F. Hutchinson, "The Community Composition of California Coastal Sage Scrub," *Vegetatio* 35 (1977): 21–33; Walter E. Westman, "Factors Influencing the Distribution of Species of Californian Coastal Sage Scrub," *Ecology* 62 (1981): 439–455; J. F. O'Leary, "Coastal Sage Scrub: Threats and Current Status," *Fremontia* 23 (1995): 27–31; Atwood and Noss, "Gnatcatchers and Development"; Clark, "NCCP"; Mantell, "Beyond Single Species"; Welner, "Natural Communities Conservation Planning"; Stevens, "Utility Participation"; Mary K. Chase et al., "Single Species as Indicators of Species Richness and Composition in California Coastal Sage Scrub Birds and Small Mammals," *CB* 14 (2000): 474–487; Jay E.

Diffendorfer et al., "Developing Terrestrial, Multi-Taxon Indices of Biological Integrity: An Example from Coastal Sage Scrub," *BC* 140 (2007): 130–141; Lauren A. Hierl et al., "Assessing and Prioritizing Ecological Communities for Monitoring in a Regional Habitat Conservation Plan," *Environmental Management* 42 (2008): 165–179; Heather L. Hulton VanTassel et al., "Environmental Change, Shifting Distributions, and Habitat Conservation Plans: A Case Study of the California Gnatcatcher," *Ecology and Evolution* 7 (2017): 10326–10338; Tarlock, "Fred Bosselman"; USFWS, "Endangered and Threatened Wildlife and Plants; Proposed Rule to List the Coastal California Gnatcatcher as Endangered," 56 Fed. Reg. 47053 (Sept. 17, 1991); Kenworthy, "Babbitt Clears Compromise"; Matt Davis, Søren Faurby, and Jens-Christian Svenning, "Mammal Diversity Will Take Millions of Years to Recover from the Current Biodiversity Crisis," *PNAS* 115 (2018): 11262–11267; Reid and Murphy, "Providing a Regional Context."

20. Craig W. Thomas, *Bureaucratic Landscapes: Interagency Cooperation and the Preservation of Biodiversity* (Cambridge, Mass.: MIT Press, 2003); Olander and Young, "Integrating Large-Scale Planning"; Peter F. Brussard and Dennis D. Murphy, "Subregionalization for Natural Communities Conservation Planning," California Department of Fish and Game, Southern California Coastal Sage Scrub Natural Communities Conservation Plan: Scientific Review Panel Conservation Guidelines and Documentation § 5 (1992); Atwood and Noss, "Gnatcatchers and Development"; Fulton, *Reluctant Metropolis*, 218.

21. Mantell, "Beyond Single Species"; California Department of Fish and Wildlife, Habitat Conservation Planning Branch, "Scientific Input," https://www.wildlife.ca.gov/Conservation/Planning/NCCP/Scientific-Input.

22. NCCP Scientific Review Panel, *Conservation Guidelines for Coastal Sage Scrub* (Sacramento: California Department of Fish and Game, 1993); Reid and Murphy, "Providing a Regional Context"; Stevens, "Utility Participation"; California Department of Fish and Game, California Resources Agency, *Southern California Coastal Sage Scrub NCCP Conservation Guidelines* (Sacramento, August 1993); Atwood and Noss, "Gnatcatchers and Development", Ebbin, "Southern California Approach"; quote from Pollak, *Future of Habitat Conservation?* 25; Jonathan L. Atwood, David R. Bontrager, and Amy L. Gorospe, "Use of Refugia by California Gnatcatchers Displaced by Habitat Loss," *WB* 29 (1998): 406–412; Sandy J. Andelman and William F. Fagan, "Umbrellas and Flagships: Efficient Conservation Surrogates or Expensive Mistakes?" *PNAS* 97 (2000): 5954–5959; William B. Miller and Clark S. Winchell, "A Comparison of Point-Count and Area-Search Surveys for Monitoring Site Occupancy of the Coastal California Gnatcatcher (*Polioptila californica californica*)," *Con-*

dor 118 (2016): 329–337; 56 Fed. Reg. 47053; Chase et al., "Single Species"; Daniel Rubinoff, "Evaluating the California Gnatcatcher as an Umbrella Species for Conservation of Southern California Coastal Sage Scrub," *CB* 15 (2001): 1374–1383; Diffendorfer et al., "Developing Terrestrial, Multi-Taxon Indices."

23. Marla Cone, "26,000 Acres Enrolled as Gnatcatcher Habitat," *LAT,* May 6, 1992; Welner, "Natural Communities Conservation Planning"; Tara L. Muller, "Natural Community Conservation Planning: Preserving Species or Developer Interests?" *Endangered Species UPDATE* 14 (1997): 26–28; Michael Jasny, Joel Reynolds, and Ann Notthoff, *Leap of Faith: Southern California's Experiment in Natural Community Conservation Planning* (Washington, D.C.: Natural Resources Defense Council, 1997); Tarlock, "Fred Bosselman"; Melinda Taylor and Holly Doremus, "Habitat Conservation Plans and Climate Change: Recommendations for Policy," *ELR* 45 (2015): 10866. See Spirit of Sage Council v. Norton, 294 F.Supp.2d 67, 72 (D.D.C. 2003), *amended sub nom.,* Spirit of Sage Council v. Norton, No. CIV.A. 98-1873, 2004 WL 1326279 (D.D.C. June 10, 2004), *and appeal dismissed, judgment vacated in part sub nom.,* Spirit of Sage Council v. Norton, 411 F.3d 225 (D.C. Cir. 2005); Wheeler and Rowberry, "Habitat Conservation Plans." Ultimately the courts sided with the USFWS and determined that the No Surprise rule did not violate either the ESA or the Administrative Procedure Act, Spirit of Sage Council v. Kempthorne, 511 F.Supp.2d 31, 46 (D.D.C. 2007); Parker, "Comment." An amendment to 50 CFR § 13.28(2)(5) on October 1, 1999, allows the USFWS to revoke an incidental take permit if the permitted activity is found to jeopardize the maintenance or recovery of a species, but take permits issued before this date cannot be revoked for that reason; Ebbin, "Southern California Approach."

24. Welner, "Natural Communities Conservation Planning"; Thomas, *Bureaucratic Landscapes;* Marla Cone and Eric Bailey, "Bird Advocates Say Tollway Unit Hurried Grading Job," *LAT,* Aug. 29, 1991; Frank Messina, "City Official Says He Let Developer Clear a Disputed Habitat," *LAT,* June 19, 1991; Maria Newman, "Clearing of Tract Allowed," *LAT,* July 27, 1991; Maria Newman, "Firm Accused of Tearing Up Bird Habitat," *LAT,* June 18, 1991; Marla Cone, "Gnatcatcher Habitat Loss Called 'Significant,'" *LAT,* Apr. 21, 1992; J. R. Reynolds, "Singing on the Brink: Can a Songbird Save Our Southern Coast?" *Land Use Forum,* Summer 1993, 229–231; Atwood and Noss, "Gnatcatchers and Development."

25. Welner, "Natural Communities Conservation Planning," "Imperiled" quotation in affidavit for Endangered Species Comm. of Bldg. Indus. Ass'n of S. Cal. v. Babbitt, 852 F.Supp. 32, 40, 41 (D.D.C. 1994), *as amended on reconsideration* (June 16, 1994) (footnote omitted) (citation omitted); Trounson and Pinsky, "Gnatcatcher";

Welner, "Natural Communities Conservation Planning"; Kripke, "Conservation Planning."

26. Mantell, "Beyond Single Species"; Welner, "Natural Communities Conservation Planning"; USFWS, "Endangered and Threatened Wildlife and Plants; Determination of Whether Designation of Critical Habitat for the Coastal California Gnatcatcher Is Prudent," 64 Fed. Reg. 5957, 5961 (Feb. 8, 1991); Pollak, *Future of Habitat Conservation?*

27. 81 Fed. Reg. 59952; *Coastal California Gnatcatcher 5-Year Review: Summary and Evaluation* (Carlsbad, Calif.: USFWS, Sept. 29, 2010), https://www.fws.gov/carlsbad/SpeciesStatusList/5YR/20100929_5YR_CAGN.pdf.

28. 81 Fed. Reg. 59952; Claudia Leyva et al., "Coastal Landscape Fragmentation by Tourism Development: Impacts and Conservation Alternatives," *Natural Areas Journal* 26 (2006): 117–125.

29. Pincetl, "Urban Fringe"; Pincetl, Watt, and Santos, "Land Use Regulation"; Camacho, Taylor, and Kelly, "Lessons"; Evans et al., "Species Recovery"; California Department of Fish and Wildlife, Habitat Conservation Planning Branch, "Natural Community Conservation Planning (NCCP)," https://www.wildlife.ca.gov/conservation/planning/nccp.

30. California Department of Fish and Wildlife, "Grants for NCCPs and HCPs," https://wildlife.ca.gov/Conservation/Planning/NCCP/Grants; Pincetl, Watt, and Santos, "Land Use Regulation"; Camacho, Taylor, and Kelly, "Lessons"; Pollak, *NCCP;* Jennifer Schlotterbeck, "Comment: Preserving Biological Diversity with Wildlife Corridors: Amending the Guidelines to the California Environmental Quality Act," *Ecology Law Quarterly* 30 (2003): 955–990; Pincetl, "Urban Fringe"; Daniel Press, *Saving Open Space: The Politics of Local Preservation in California* (Berkeley: University of California Press, 2002).

31. Feldman and Jonas, "Sage Scrub Revolution"; Camacho, Taylor, and Kelly, "Lessons"; Taylor and Doremus, "Habitat Conservation Plans"; Alejandro E. Camacho et al., "Improving Emerging Regulatory Experiments in Permit Process Coordination for Endangered Species and Aquatic Resources in California," *ELR* 46 (2016): 10131–10140; Martin Landau, "Redundancy, Rationality, and the Problem of Duplication and Overlap," *Public Administration Review* 29 (1969): 346–358.

32. Press, *Saving Open Space,* 65; Pincetl, "Urban Fringe"; Kripke, "Conservation Planning"; Pollak, *Future of Habitat Conservation?* table 1.

33. Kripke, "Conservation Planning."

34. Pollak, *Future of Habitat Conservation?;* California Department of Fish and Wildlife, "NCCP–Plan Summary–County of Orange (Central/Coastal) NCCP/

HCP," https://www.wildlife.ca.gov/Conservation/Planning/NCCP/Plans/Orange -Coastal. The Santa Margarita Company is now called the Rancho Mission Viejo Company.

35. Pollak, *Future of Habitat Conservation?*; Kripke, "Conservation Planning"; Stevens, "Utility Participation"; Marla Cone, "Gnatcatcher Study Rebuts Finding That It's Imperiled," *LAT,* July 24, 1991; Deborah Schoch, "Songbird of Peace: Plan to Save Gnatcatcher, Other Wildlife May Be Model for Protecting Environment, Jobs," *LAT,* Apr. 14, 1996.

36. Natural Communities Coalition, "About NCC," https://occonservation.org /about-ncc/; Natural Communities Coalition, *Nature Reserve of Orange County, County of Orange Central/Coastal NCCP/HCP, 2017 Annual Report,* https://occonservation .org/wp-content/uploads/2018/11/2017-Annual-Report-v2-reduced-size.pdf; Pollak, *Future of Habitat Conservation?;* Stevens, "Utility Participation." The Irvine Company's investment in the NCCP preserve network was explained by the company's chairman Donald Bren in 2001: "[It is] the large-scale protection of wilderness lands it provides in close proximity to urban life where it can be more readily enjoyed"; cited in Kripke, "Conservation Planning"; "The Irvine Company to Contribute More Than 11,000 Acres to Irvine Ranch Land Reserve as Permanent Open Space" (Press release, the Irvine Company, Newport Beach, Calif., Nov. 28, 2001); Natural Communities Coalition, *Nature Reserve of Orange County, County of Orange Central/ Coastal/NCCP/HCP, 2015 Annual Report,* https://occonservation.org/wp-content/up loads/mdocs/2015annualreport.pdf.

37. Orange County Transportation Authority NCCP/HCP, NCCP Permit 2835-2017-001-05 document, June 2017, 24, 48.

38. Orange County Transportation Authority NCCP/HCP, NCCP Permit 2835-2017-001-05 document, June 2017, 2, 4–5; Friends of Harbors, Beaches, and Parks was formed in 1997 by a group of environmentalists to coordinate open space management and environmental education in the wake of Orange County's bankruptcy and reorganization in 1996–1997; see Friends of Harbors, Beaches, and Parks, "About Us," https://www.fhbp.org/about-us/history/; Orange County Transportation Authority, "M2 Natural Community Conservation Plan/Habitat Conservation Plan–First Annual Report," 2019, https://www.octa.net/pdf/OCTA_M2_NCCP_HCP _FirstAnnualReport_Final_Jul19v5.pdf.

39. Pulaski v. Chrisman, 352 F.Supp.2d 1105, 1113 (C.D. Ca. 2005), *aff'd,* 127 F.App'x 993 (9th Cir. 2005); David Reyes, "A Farewell Wave to Beach Life," *LAT,* Feb. 27, 2006.

40. Kevin B. Clark, *El Toro Conservation Area Post-Fire Assessment,* Prepared for the Nature Reserve of Orange County (San Diego: Clark Biological Services, 2013, doi.org/10.13140/RG.2.2.16220.23686); Kevin B. Clark, *El Toro Conservation Area Resource Management Plan,* Prepared for the Nature Reserve of Orange County (San Diego: Clark Biological Services, 2007, doi.org/10.13140/RG.2.2.24608.84482); Pat Brennan, "Deal for Nature Reserve Ends after 15 Years," *Orange County Register,* May 22, 2011; Dr. Milan Mitrovich, the Nature Reserve of Orange County, pers. comm., May 28, 2019.

41. Michelle Ouellette and Charles Landry, "The Western Riverside County Multiple Species Habitat Conservation Plan: Looking Forward after Ten Years," *Natural Resources and Environment* 29 (2015): 40–46; Charlie Landry, "Wildlife Crossings and Connectivity within the Western Riverside County MSHCP," presentation at the 2019 National HCP Coalition annual meeting in Shepherdstown, W.Va., Nov. 13–15, 2019, https://www.nhcpcoalition.org/2019-annual-meeting/; Xiongwen Chen et al., "Spatial Structure of Multispecies Distributions in Southern California, USA," *BC* 124 (2005): 169–175; Western Riverside County Regional Conservation Authority, "Reserves," http://www.wrc-rca.org/habitat-conservation/reserves/; Pincetl, "Urban Fringe."

42. Camacho, Taylor, and Kelly, "Lessons."

43. Columbia University Integrative Capstone Workshop in Sustainability Management, "Understanding the Species Mitigation Market in the United States," Prepared for the Nature Conservancy, faculty advisor: George Sarrinikolaou (New York: Columbia University, 2016); Ouellette and Landry, "Western Riverside." This includes loans via the Water Resources Reform and Development Act of 2014 (Pub. L. No. 113-121, 128 Stat. 1193), which includes funding for habitat conservation.

44. Hulton VanTassel et al., "Environmental Change."

45. Davis, "High Noon"; Western Riverside County MSHCP Signed Implementing Agreement, RVPUB\MO\655487 (2003), https://rctlma.org/Portals/0/mshcp/volume3/Implementing_Agree.pdf, 18–19.

46. Scott A. Morrison and Walter M. Boyce, "Conserving Connectivity: Some Lessons from Mountain Lions in Southern California," *CB* 23 (2009): 275–285; Paulek v. W. Riverside Cty. Reg'l Conservation Auth., 238 Cal.App.4th 583, 190 Cal. Rptr.3d 59, 64 (2015), *as modified on denial of reh'g* (July 17, 2015).

47. See Paulek, 190 Cal.Rptr.3d at 66, https://www.courts.ca.gov/opinions/archive/E059133M.PDF; Western Riverside County MSHCP Signed Implementing Agreement, RVPUB\MO\655487.

48. See Paulek, 190 Cal.Rptr.3d at 66.

49. See Sw. Ctr. for Biological Diversity v. Bartel, 470 F.Supp.2d 1118, 1122–23 (S.D. Cal. 2006), *appeal dismissed and remanded,* 409 F.App'x 143 (9th Cir. 2011).

50. Resolution No. 2015-011, Resolution of the Board of Directors of the Western Riverside County Regional Conservation Authority to vacate and set aside Resolution No. 12–002 certifying and approving the Warm Springs criteria refinement, and to rescind the notice of exemption concerning the criteria refinement, http://www.wrc-rca.org/archivecdn/Permit_Docs/Resolutions/Resolution%20No.%2015–011%20Rescind%2012-002%20Warm%20Springs.pdf; Brian Beck, Western Riverside County Regional Conservation Authority, pers. comm., May 14, 2019.

51. Pollak, *NCCP;* Pollak, *Future of Habitat Conservation?;* Camacho, Taylor, and Kelly, "Lessons."

52. You can see maps of these highly irregular planning areas here: https://nrm.dfg.ca.gov/FileHandler.ashx?DocumentID=68626&inlineandhttps://www.sandiegocounty.gov/content/dam/sdc/pds/mscp/docs/mscp_areas.pdf; Reed et al., *Wildlife Response to Human Recreation.*

53. Pollak, *NCCP;* Pollak, *Future of Habitat Conservation?* table 1; Ebbin, "Southern California Approach"; Camacho, Taylor, and Kelly, "Lessons"; Buse, "Multi-Species Habitat."

54. Rob Davis, "Once a National Model, Habitat Plan Faces Uncertain Future," *Voice of San Diego,* Apr. 16, 2007; Buse, "Multi-Species Habitat"; Weyerhaeuser Co. v. U.S. Fish & Wildlife Serv., 139 S.Ct. 361 (2018); Rahn, Doremus, and Diffendorfer, "Species Coverage."

55. Wikipedia's entry on the Palos Verdes Peninsula; J. Phillips, *Palos Verdes Estates* (Charleston, S.C.: Arcadia, 2010); Rancho Palos Verdes Natural Community Conservation Plan and Habitat Conservation Plan, Final Draft, https://www.rpvca.gov/DocumentCenter/View/13211/NCCPHCP, May 2018.

56. Rancho Palos Verdes Natural Community Conservation Plan and Habitat Conservation Plan, 29, 40; USFWS, "Draft City of Rancho Palos Verdes Natural Community Conservation Plan and Habitat Conservation Plan and Draft Environmental Assessment, City of Rancho Palos Verdes, Los Angeles County, California," 84 Fed. Reg. 13308 (Apr. 4, 2019).

57. Rancho Palos Verdes Natural Community Conservation Plan and Habitat Conservation Plan; California Department of Fish and Wildlife, "Wildlife Conservation Board Funds Environmental Improvements and Acquisition Projects," http://www.news.ca.gov, Nov. 21, 2013; Pincetl, "Urban Fringe"; Mike Davis, *Ecology of Fear: Los Angeles and the Imagination of Disaster* (New York: Vintage Books, 1998), 85.

Chapter 7. Is the NCCP Policy a Success?

1. Atwood, "Gnatcatchers"; Amy G. Vandergast et al., "Distinguishing Recent Dispersal from Historical Genetic Connectivity in the Coastal California Gnatcatcher," *Scientific Reports* 9 (2019): art1355, https://doi.org/10.1038/s41598-018-37712-2, p. 7.

2. Kelly R. Barr et al., "Habitat Fragmentation in Coastal Southern California Disrupts Genetic Connectivity in the Cactus Wren (*Campylorhynchus brunneicapillus*)," *Molecular Ecology* 24 (2015): 2349–2363. The tricolor blackbird (*Agelaius tricolor*) was under review for listing in 2019; U.S. Department of Interior, Office of the Inspector General, ed., "Investigative Report: On Allegations against Julie MacDonald Deputy Assistant Secretary, Fish, Wildlife and Parks" (Washington, D.C.: U.S. Department of the Interior, 2007); Joe Roman, *Listed: Dispatches from America's Endangered Species Act* (Cambridge, Mass.: Harvard University Press, 2011), 73–75; Emily E. Puckett, Dylan C. Kesler, and D. Noah Greenwald, "Taxa, Petitioning Agency, and Lawsuits Affect Time Spent Awaiting Listing under the U.S. Endangered Species Act," *BC* 201 (2016): 220–229; Dino Grandoni, "The Energy 202: Trump Administration Slow to Declare Species Endangered Amid Extinction Crisis," *Washington Post,* May 7, 2019.

3. USFWS, "Quino Checkerspot Butterfly (*Euphydryas editha quino*) 5-Year Review" (Carlsbad, Calif., Aug. 13, 2009), https://www.fws.gov/carlsbad/SpeciesStatus List/5YR/20090813_5YR_QCB.pdf; Camille Parmesan et al., "Endangered Quino Checkerspot Butterfly and Climate Change: Short-Term Success but Long-Term Vulnerability?" *Journal of Insect Conservation* 19 (2014): 185–204; Daniel Pollak, *The Future of Habitat Conservation? The NCCP Experience in Southern California* (Sacramento: California Research Bureau CRB-01-009, 2001); California Department of Fish and Wildlife, "NCCP Plan Summary–San Diego Multiple Species Conservation Program (MSCP)," https://www.wildlife.ca.gov/Conservation/Planning/NCCP /Plans/San-Diego-MSCP; USFWS, "Endangered and Threatened Wildlife and Plants; Emergency Rule to List the Pacific Pocket Mouse as Endangered," 59 Fed. Reg. 5306 (Feb. 3, 1994).

4. USFWS, "Endangered and Threatened Wildlife and Plants; Special Rule Concerning Take of the Threatened Coastal California Gnatcatcher," 58 Fed. Reg. 65088 (Dec. 10, 1993); Christine M. Ryan, John Wilson, and William Fulton, "Living on the Edge: Growth Policy Choices for Ventura County," in *Up against the Sprawl: Public Policy and the Making of Southern California,* ed. Jennifer Wolch, Manuel Pastor Jr., and Peter Dreier (Minneapolis: University of Minnesota Press, 2004), 329–330.

5. K. Smallwood et al., "Indicators Assessment for Habitat Conservation Plan of Yolo County, California, USA," *Environmental Management* 22 (1998): 947–958;

Richard A. Redak, "Arthropods and Multispecies Habitat Conservation Plans: Are We Missing Something?" *Environmental Management* 26 (2000): S97–S107; P. Dee Boersma et al., "How Good Are Endangered Species Recovery Plans?" *BioScience* 51 (2001): 643–649; Martin F. J. Taylor, Kierán F. Suckling, and Jeffrey J. Rachlinski, "The Effectiveness of the Endangered Species Act: A Quantitative Analysis," *Bio-Science* 55 (2005): 360–367; Matthew E. Rahn, Holly Doremus, and James Diffen-dorfer, "Species Coverage in Multispecies Habitat Conservation Plans: Where's the Science?" *BioScience* 56 (2006): 614; Christian Langpap and Joe Kerkvliet, "Endangered Species Conservation on Private Land: Assessing the Effectiveness of Habitat Conservation Plans," *Journal of Environmental Economics and Management* 64 (2012): 1–15; John Buse, "Can a Multi-Species Habitat Conservation Plan Save San Diego's Vulnerable Vernal Pool Species?" *Golden Gate University Environmental Law Journal* 6 (2012): 77; Mary K. Chase et al., "Single Species as Indicators of Species Richness and Composition in California Coastal Sage Scrub Birds and Small Mammals," *CB* 14 (2000): 474–487; Thomas A. Scott and James E. Sullivan, "The Selection and De-sign of Multiple-Species Habitat Preserves," *Environmental Management* 26 (2000): S37–S53; Redak, "Arthropods."

6. Pollak, *Future of Habitat Conservation?*

7. Jay E. Diffendorfer et al., "Developing Terrestrial, Multi-Taxon Indices of Bio-logical Integrity: An Example from Coastal Sage Scrub," *BC* 140 (2007): 130–141; David B. Lindenmayer, Maxine P. Piggott, and Brendan A. Wintle, "Counting the Books while the Library Burns: Why Conservation Monitoring Programs Need a Plan for Action," *Frontiers in Ecology and the Environment* 11 (2013): 549–555; Court-ney L. Larson et al., "Accessibility Drives Species Exposure to Recreation in Frag-mented Urban Reserve Network," *Landscape and Urban Planning* 175 (2018): 62–71; Milan Mitrovich et al., "Balancing Conservation and Recreation," *California Fish and Wildlife Journal, Recreation Special Issue* (2020): 11–28; Jonathan L. Atwood and David R. Bontrager, "California Gnatcatcher (*Polioptila californica*)," version 2.0, in *The Birds of North America*, 2001, ed. A. F. Poole and F. B. Gill (Cornell Lab of Orni-thology, Ithaca, N.Y.), https://doi.org/10.2173/bna.574.

8. Lydia P. Olander and Ben L. Young, "Integrating Large-Scale Planning into En-vironmental Markets and Related Programs: Status and Trends," NI WP 17-03 (Dur-ham, N.C.: Duke University, Nicholas School of the Environment, 2016).

9. Pollak, *Future of Habitat Conservation?;* Columbia University Integrative Cap-stone Workshop in Sustainability Management, "Understanding the Species Miti-gation Market in the United States," Prepared for the Nature Conservancy, faculty advisor: George Sarrinikolaou (New York: Columbia University, 2016); "Packard

Foundation Completes Five-Year California Conservation Initiative," *Philanthropy News Digest,* Nov. 14, 2003, https://philanthropynewsdigest.org/news/packard-foundation-completes-five-year-california-conservation-initiative.

10. Richard Conniff, "Why Isn't Publicly Funded Conservation on Private Land More Accountable?" *Yale E360,* July 23, 2019, https://e360.yale.edu/features/why-isnt-publicly-funded-conservation-on-private-land-more-accountable.

11. Pollak, *Future of Habitat Conservation?;* Stephanie Pincetl, "The Preservation of Nature at the Urban Fringe," in Wolch, Pastor, and Dreier, *Up against the Sprawl,* 225–251; Ryan, Wilson, and Fulton, "Living on the Edge."

12. William Fulton, *The Reluctant Metropolis: The Politics of Urban Growth in Los Angeles* (Baltimore: Johns Hopkins University Press, 2001); Pincetl, "Preservation of Nature"; Joe Nelson, "Former Top San Bernardino County Official, Acquitted in Public Corruption Trial, Files $35 Million Claim against County and State," *Sun* (San Bernardino), Dec. 18, 2017; USFWS, "Revised Designation of Critical Habitat for the Coastal California Gnatcatcher (*Polioptila californica californica*); Final Rule," 72 Fed. Reg. 72010, 72021–72022 (Dec. 19, 2007); USFWS, "Proposed Upper Santa Ana River Habitat Conservation Plan and Draft Environmental Impact Statement; San Bernardino County, CA," 84 Fed. Reg. 67292 (Dec. 9, 2019); USFWS, "Final Environmental Impact Statement for the Upper Santa Ana River Wash Habitat Conservation Plan; San Bernardino County, CA," 85 Fed. Reg. 29474 (May 15, 2020); Upper Santa Ana River Habitat Conservation Plan, http://www.uppersarhcp.com/About.aspx.

13. Ethan Elkind, "Climate Actions for California: Recommendations for Governor Brown's Final Term," *Center for Law, Energy and the Environment Publications* 8 (Berkeley: University of California, December 2015); Marty Graham, "One Million Dollar Houses Atop San Pasqual Valley," *San Diego Reader,* May 21, 2018.

14. San Pasqual Valley Preservation Alliance: https://www.spvpa.org/groups/; see San Pasqual Valley Preservation Alliance's archive of responses to the project's Draft Environmental Impact Report: https://www.spvpa.org/safari-highlands-ranch/#announcements; San Pasqual Valley Preservation Alliance, "Harvest Hills?" https://www.spvpa.org/harvest-hills/.

15. Pincetl, "Transforming"; Craig W. Thomas, *Bureaucratic Landscapes: Interagency Cooperation and the Preservation of Biodiversity* (Cambridge, Mass.: MIT Press, 2003); Greg Hise and William Deverell, "Introduction: The Metropolitan Nature of Los Angeles," in *Land of Sunshine: An Environmental History of Metropolitan Los Angeles,* ed. Deverell and Hise (Pittsburgh: University of Pittsburgh Press, 2005), 1–12; Maria João Santos, Terry Watt, and Stephanie Pincetl, "The Push and Pull of Land

Use Policy: Reconstructing 150 Years of Development and Conservation Land Acquisition," *PLoS ONE* 9 (2014): https://doi.org/10.1371/journal.pone.0103489; Elkind, "Climate Actions"; Mike Davis, *Ecology of Fear: Los Angeles and the Imagination of Disaster* (New York: Vintage Books, 1998), 62; Jared Orsi, "Flood Control Engineering in the Urban Ecosystem," in Deverell and Hise, *Land of Sunshine,* 135–151; City of Los Angeles, Los Angeles River Revitalization, https://www.lariver.org.

16. Pollak, *Future of Habitat Conservation?*

17. Andrea Olive, "It Is Just Not Fair: The Endangered Species Act in the United States and Ontario," *Ecology and Society* 21 (2016): https://doi.org/10.5751/ES-086 27-210313; Richard E. Matland, "Synthesizing the Implementation Literature: The Ambiguity-Conflict Model of Policy Implementation," *Journal of Public Administration Research and Theory* 5 (1995): 145–174. Matland describes "high ambiguity/high conflict" policies as "symbolic implementation," since much of the conflict stems over "highly salient symbols" that emphasize differences among stakeholders (in the NCCP's case, species extinction versus private property rights).

18. A. Dan Tarlock, "Fred Bosselman as Participant-Observer Lawyer: The Case of Habitat Conservation Planning," *Florida State University Journal of Land Use and Environmental Law* 17 (2018): 43–55; Pollak, *Future of Habitat Conservation?* (MSCP project manager for SANDAG on 61). For the importance of adaptive capacity in environmental laws, see Ahjond Garmestani et al., "Untapped Capacity for Resilience in Environmental Law," *PNAS* 116 (2019): 19899–19904.

19. DeAnne Parker, "Comment: Natural Community Conservation Planning: California's Emerging Ecosystem Management Alternative," *University of Baltimore Journal of Environmental Law* 6 (1997): 107–140. A 1999 amendment to the ESA allows USFWS to revoke take permits (approved after Oct. 1, 1999) if a species "declines to the extent that continuation of the permitted activity would be detrimental to the maintenance or recovery of the affected population"; 50 CFR § 13.28(a)(5).

20. Allan McConnell, "Policy Success, Policy Failure and Gray Areas In-Between," *Journal of Public Policy* 30 (2010): 345–362.

21. Pollak, *Future of Habitat Conservation?* 62.

22. Although see Barr et al., "Habitat Fragmentation": populations of cactus wrens on Camp Pendleton may be isolated from other populations. Erin Conlisk et al., "Predicting the Impact of Fire on a Vulnerable Multi-Species Community Using a Dynamic Vegetation Model," *Ecological Modelling* 301 (2015): 27–39; Zachary Principe et al., *50-Year Climate Scenarios and Plant Species Distribution Forecasts for Setting Conservation Priorities in Southwestern California* (San Francisco: Nature Conservancy of California, 2013).

23. The California Department of Fish and Wildlife assumes the NCCP is a conservation policy that "is designed to facilitate the adaptation of wildlife to climate change" owing to the NCCP's emphasis of landscape-scale connected reserves: https://wildlife.ca.gov/Conservation/Climate-Science/Case-Studies/NCCP; Elkind, "Climate Actions"; Melinda Taylor and Holly Doremus, "Habitat Conservation Plans and Climate Change: Recommendations for Policy," *ELR* 45 (2015): 10866. The exclusion of climate change impacts is not unique to the NCCP: John Kostyack et al., "Beyond Reserves and Corridors: Policy Solutions to Facilitate the Movement of Plants and Animals in a Changing Climate," *BioScience* 61 (2011): 713–719; D. Richard Cameron et al., "Ecosystem Management and Land Conservation Can Substantially Contribute to California's Climate Mitigation Goals," *PNAS* 114 (2017): 12833–12838; Frank W. Davis and Elizabeth A. Chornesky, "Adapting to Climate Change in California," *Bulletin of the Atomic Scientists* 70 (2014): 62–73; Adina M. Merenlender et al., "Stewardship, Conservation, and Restoration in the Context of Environmental Change," in *Ecosystems of California,* ed. Harold Mooney and Erika Zavaleta (Oakland: University of California Press, 2016), 925–944; Principe et al., *50-Year Climate Scenarios.*

24. Merenlender et al., "Stewardship, Conservation, and Restoration"; Peter S. Alagona et al., "Population and Land Use," in Mooney and Zavaleta, *Ecosystems of California,* 75–94; Diane E. Pataki et al., "Urban Ecosystems," in Mooney and Zavaleta, *Ecosystems of California,* 885–898; Shelby Grad, "Los Angeles Hits a Milestone: 4 Million People and Counting," *LAT,* May 2, 2017; Alagona et al., "Population and Land Use"; Andrew Khouri, "One Solution to Southern California's Housing Crisis: Building in Tight Spaces, Small Lots," *LAT,* Dec. 23, 2016; Steve Lopez, "Everyone Loves L.A.—and That's the Problem," *LAT,* June 17, 2017; Rutherford H. Platt, *Land Use and Society: Geography, Law, and Public Policy,* 3rd ed. (Washington, D.C.: Island Press, 2014); California Department of Fish and Wildlife, "Summary of Natural Community Conservation Plans (NCCPs)," October 2017, https://nrm.dfg.ca.gov /FileHandler.ashx?DocumentID=15329&inline.

25. Norbert Müller and Peter Werner, "Urban Biodiversity and the Case for Implementing the Convention on Biological Diversity in Towns and Cities," in *Urban Biodiversity and Design,* ed. Müller, Werner, and John G. Kelcey (Oxford: Wiley-Blackwell, 2010); Marina Alberti, "Eco-Evolutionary Dynamics in an Urbanizing Planet," *Trends in Ecology and Evolution* 30 (2015): 114–126; Mark J. McDonnell and Amy K. Hahs, "Adaptation and Adaptedness of Organisms to Urban Environments," *Annual Review of Ecology, Evolution, and Systematics* 46 (2015): 261–280; Ken A. Thompson, Marie Renaudin, and Marc T. J. Johnson, "Urbanization Drives

the Evolution of Parallel Clines in Plant Populations," *PRSB* 283 (2016): https://
doi.org/10.1098/rspb.2016.2180; Marine Alberti et al., "Global Urban Signatures of
Phenotypic Change in Animal and Plant Populations," *PNAS* 114 (2017): 8951–8956;
Dolph Schluter and Matthew W. Pennell, "Speciation Gradients and the Distribution
of Biodiversity," *Nature* 546 (2017): 48–55; M. T. J. Johnson and J. Munshi-South,
"Evolution of Life in Urban Environments," *Science* 358 (2017): eaam8327; Constan-
tino Macías Garcia et al., "Becoming Citizens: Avian Adaptations to Urban Life," in
Ecology and Conservation of Birds in Urban Environments, ed. Enrique Murgui and
Marcus Hedblom (Cham, Switzerland: Springer, 2017), 91–112; Kristin M. Winchell
et al., "Phenotypic Shifts in Urban Areas in the Tropical Lizard *Anolis cristatellus*,"
Evolution 70 (2016): 1009–1022; Kristin M. Winchell et al., "Linking Locomotor
Performance to Morphological Shifts in Urban Lizards," *PRSB* 285 (2018): 20180229;
V. K. Tokhtari and R. Wittig, "Evolution and Development of Plant Populations in
Technogenous Ecotopes," *Soil Science* 1 (2001): 97–105; Müller and Werner, "Urban
Biodiversity"; V. K. Tokhtar, Yu. K. Vinogradova, and A. S. Groshenko, "Micro-
evolution and Invasiveness of *Oenothera* L. species (subsect. *Oenothera,* Onagra-
ceae) in Europe," *Russian Journal of Biological Invasions* 2 (2011): 273–280. Evolving
urban adaptation traits in Europe, then spreading throughout the world via Euro-
pean colonization, is a popular strategy for plants: see Frank A. La Sorte, Michael L.
McKinney, and Petr Pyšek, "Compositional Similarity among Urban Floras within
and across Continents: Biogeographical Consequences of Human-Mediated Biotic Interchange," *Global Change Biology* 13 (2007): 913–921; Frank A. La Sorte and
Petr Pyšek, "Extra-Regional Residence Time as a Correlate of Plant Invasiveness:
European Archaeophytes in North America," *Ecology* 90 (2009): 2589–2597; Ingo
Kowarik, "Novel Urban Ecosystems, Biodiversity, and Conservation," *Environmental Pollution* 159 (2011): 1974–1983; Richard J. Hobbs, Eric S. Higgs, and Carol Hall,
eds., *Novel Ecosystems: Intervening in the New Ecological World Order* (Oxford: Wiley-
Blackwell, 2013); Diane E. Pataki et al., "Urban Ecosystems," in Mooney and Zavaleta, *Ecosystems of California,* 885–898; Richard J. Hobbs, Eric Higgs, and James A.
Harris, "Novel Ecosystems: Implications for Conservation and Restoration," *Trends
in Ecology and Evolution* 24 (2009): 599–605; Frank A. La Sorte et al., "The Phylogenetic and Functional Diversity of Regional Breeding Bird Assemblages Is Reduced
and Constricted through Urbanization," *Diversity and Distributions* 24 (2018): 928–
938; Carolina Murcia et al., "A Critique of the 'Novel Ecosystem' Concept," *Trends in
Ecology and Evolution* 29 (2014): 548–553; Sebastián Martinuzzi et al., "Future Land-
Use Changes and the Potential for Novelty in Ecosystems of the United States," *Ecosystems* 18 (2015): 1332–1342.

26. Stephen T. Garnett and Les Christidis, "Taxonomy Anarchy Hampers Conservation," *Nature* 546 (2017): 25–27; Rachel Cernansky, "Biodiversity Moves beyond Counting Species," *Nature* 546 (2017): 22–24; Forest Isbell et al., "Linking the Influence and Dependence of People on Biodiversity across Scales," *Nature* 546 (2017): 65–72; Laura J. Pollock, Wilfried Thuiller, and Walter Jetz, "Large Conservation Gains Possible for Global Biodiversity Facets," *Nature* 546 (2017): 141–144; USFWS, "Endangered and Threatened Wildlife and Plants; Endangered Species Act Compensatory Mitigation Policy," 83 Fed. Reg. 36469 (July 30, 2018), summary point A; Executive Office of the President, "Memorandum on Mitigating Impacts on Natural Resources from Development and Encouraging Related Private Investment," 80 Fed. Reg. 68743 (Nov. 3, 2015); USFWS, "U.S. Fish and Wildlife Service Mitigation Policy," 81 Fed. Reg. 83440 (Nov. 21, 2016).

27. Weyerhaeuser Co. v. U.S. Fish & Wildlife Serv., 139 S.Ct. 361 (2018); Robert Barnes, "Supreme Court Deals a Setback to the Endangered Dusky Gopher Frog," *Washington Post,* Nov. 27, 2018.

28. Jordan S. Rosenfeld, "Functional Redundancy in Ecology and Conservation," *Oikos* 98 (2002): 156–162; Owen L. Petchey and Kevin J. Gaston, "Functional Diversity: Back to Basics and Looking Forward," *Ecology Letters* 9 (2006): 741–758; William J. Ripple and Robert L. Beschta, "Trophic Cascades in Yellowstone: The First 15 Years after Wolf Reintroduction," *BC* 145 (2012): 205–213; Amy G. Vandergast et al., "Are Hotspots of Evolutionary Potential Adequately Protected in Southern California?" *BC* 141 (2008): 1648–1664; Schluter and Pennell, "Speciation Gradients."

29. SANDAG's *TransNet* Environmental Mitigation Program, https://www.sandag.org/index.asp?newsid=740&fuseaction=news.detail; Alejandro E. Camacho, Elizabeth Taylor, and Melissa Kelly, "Lessons from Area-Wide, Multi-Agency Habitat Conservation Plans in California," *ELR* 46 (2016): 10222–10248; California Department of Fish and Wildlife, Regional Advance Mitigation, https://www.wildlife.ca.gov/Conservation/Planning/Regional-Advance-Mitigation; California Department of Fish and Wildlife, Regional Conservation Investment Strategies Program, https://wildlife.ca.gov/conservation/planning/regional-conservation.

30. Camacho, Taylor, and Kelly, "Lessons"; National Habitat Conservation Plan Coalition, https://www.nhcpcoalition.org/.

31. Kendall R. Jones et al., "One-Third of Global Protected Land Is under Intense Human Pressure," *Science* 360 6390 (2018): 788–791; Rachel E. G. Kroner et al., "The Uncertain Future of Protected Lands and Waters," *Science* 364 (2019): 881–886; Piero Visconti et al., "Protected Area Targets Post-2020," *Science* 364 (2019): 239–241;

Daniel Press, *Saving Open Space: The Politics of Local Preservation in California* (Berkeley: University of California Press, 2002).

32. Sarah A. Bekessy et al., "Transparent Planning for Biodiversity and Development in the Urban Fringe," *Landscape and Urban Planning* 108 (2012): 140–149; Lyle E. Ground, Rob Slotow, and Jayanti Ray-Mukherjee, "The Value of Urban and Peri-Urban Conservation Efforts within a Global Biodiversity Hotspot," *Bothalia-African Biodiversity and Conservation* 46 (2016): doi.org/10.4102/abc.v46i2.2106; Kylie Soanes et al., "Correcting Common Misconceptions to Inspire Conservation Action in Urban Environments," *CB* 33, no. 2 (2019): 300–306; Press, *Saving Open Space,* 144; Santos, Watt, and Pincetl, "Push and Pull"; Pincetl, Watt, and Santos, "Land Use Regulation"; Pincetl, "Preservation of Nature," 230; Robert Gottlieb, *Reinventing Los Angeles: Nature and Community in the Global City* (Cambridge, Mass.: MIT Press, 2007), 35; Laura J. Martin et al., "Conservation Opportunities across the World's Anthromes," *Diversity and Distributions* 20 (2014): 745–755; Richard Weller, "The City Is Not an Egg: Western Urbanization in Relation to Changing Conceptions of Nature," in *Nature and Cities: The Ecological Imperative in Urban Design and Planning,* ed. Frederick R. Steiner, George F. Thompson, and Armando Carbonell (Cambridge, Mass.: Lincoln Institute of Land Policy, 2016), 31–49; Visconti et al., "Protected Area Targets"; Bernie Tershy et al., "Biodiversity," in Mooney and Zavaleta, *Ecosystems of California,* 187–212; Convention on Biological Diversity, "Strategic Plan for Biodiversity 2011–2020, including Aichi Biodiversity Targets," 2010, https://www.cbd.int/sp; Eric Dinerstein et al., "An Ecoregion-Based Approach to Protecting Half the Terrestrial Realm," *BioScience* 67 (2017): 534–545; Joshua R. Ginsberg, "When Protected Areas Prove Insufficient: Cheetah and 'Protection-Reliant' Species," *PNAS* 114 (2017): 430–431; Bruno A. Walther and Lionel H. Pirsig, "Determining Conservation Priority Areas for Palearctic Passerine Migrant Birds in Sub-Saharan Africa," *Avian Conservation and Ecology* 12 (2017): https://doi.org/10.5751/ACE-00934-120102; Evan R. Buechley et al., "Identifying Critical Migratory Bottlenecks and High-Use Areas for an Endangered Migratory Soaring Bird across Three Continents," *Journal of Avian Biology* 49 (2018): https://doi.org/10.1111/jav.01629.

Chapter 8. Concrete Jungles and Granite Gardens

Note to epigraph: Anne Whiston Spirn, *The Granite Garden: Urban Nature and Human Design* (New York: Basic Books, 1984), 4.

1. Ingolf Kühn, Roland Brandl, and Stefan Klotz, "The Flora of German Cities Is Naturally Species Rich," *Evolutionary Ecology Research* 6 (2004): 749–794; Richard P. Cincotta, Jennifer Wisnewski, and Robert Engelman, "Human Population in the

Biodiversity Hotspots," *Nature* 404 (2000): 990–992; Gary W. Luck, "A Review of the Relationships between Human Population Density and Biodiversity," *Biological Review* 82 (2007): 607–645; Myla F. J. Aronson et al., "A Global Analysis of the Impacts of Urbanization on Bird and Plant Diversity Reveals Key Anthropogenic Drivers," *PRSB* 281 (2014): https://doi.org/10.1098/rspb.2013.3330; but this might be due to sampling effort and scale: see Gary W. Luck, "Why Is Species Richness Often Higher in More Densely Populated Regions?" *Animal Conservation* 13 (2010): 442–443; Christopher D. Ives et al., "Cities Are Hotspots for Threatened Species," *Global Ecology and Biogeography* 25 (2016): 117–126; Brian Czech, Paul R. Krausman, and Patrick K. Devers, "Economic Associations among Causes of Species Endangerment in the United States," *BioScience* 50 (2000): 593–601; Robert I. Mcdonald, Peter Kareiva, and Richard T. T. Forman, "The Implications of Current and Future Urbanization for Global Protected Areas and Biodiversity Conservation," *BC* 141 (2008): 1695–1703; Grant Daniels and Jamie Kirkpatrick, "Ecology and Conservation of Australian Urban and Exurban Avifauna," in *Ecology and Conservation of Birds in Urban Environments,* ed. Enrique Murgui and Marcus Hedblom (Cham, Switzerland: Springer, 2017), 343–370; Norbert Müller and Peter Werner, "Urban Biodiversity and the Case for Implementing the Convention on Biological Diversity in Towns and Cities," in *Urban Biodiversity and Design,* ed. Müller, Werner, and John G. Kelcey (Oxford: Wiley-Blackwell, 2010), 3–33; Marcus Hedblom and Enrique Murgui, "Urban Bird Research in a Global Perspective," in Murgui and Hedblom, *Ecology and Conservation of Birds,* 3–11; Michael L. McKinney, "Do Human Activities Raise Species Richness? Contrasting Patterns in United States Plants and Fishes," *Global Ecology and Biogeography* 11 (2002): 343–348; Kevin J. Gaston, "Biodiversity and Extinction: Species and People," *Progress in Physical Geography* 29 (2005): 239–247; Marco Pautasso and Michael L. McKinney, "The Botanist Effect Revisited: Plant Species Richness, County Area, and Human Population Size in the United States," *CB* 21 (2007): 1333–1340; Ingo Kowarik, "Novel Urban Ecosystems, Biodiversity, and Conservation," *Environmental Pollution* 159 (2011): 1974–1983; Diane E. Pataki et al., "Urban Ecosystems," in *Ecosystems of California,* ed. Harold Mooney and Erika Zavaleta (Oakland: University of California Press, 2016), 885–898.

2. Donald C. Dearborn and Salit Kark, "Motivations for Conserving Urban Biodiversity," *CB* 24 (2010): 432–440; Mark J. McDonnell and Amy K. Hahs, "The Future of Urban Biodiversity Research: Moving beyond the 'Low-Hanging Fruit,'" *Urban Ecosystems* 16 (2013): 397–409; Charles H. Nilon et al., "Planning for the Future of Urban Biodiversity: A Global Review of City-Scale Initiatives," *BioScience* 67 (2017): 332–342; Annenberg quoted in Bettina Boxall, "To Help Cougars Cross

Busy 101 Freeway, Annenberg Foundation Promises to Match Donations for Bridge,"
LAT, Oct. 19, 2016.

3. Richard T. T. Forman, *Towns, Ecology, and the Land* (Cambridge: Cambridge University Press, 2019), 76; Lara O'Sullivan et al., "Deforestation, Mosquitoes, and Ancient Rome: Lessons for Today," *BioScience* 58 (2008): 756–760; Daniel J. Tregidgo et al., "Rainforest Metropolis Casts 1,000-km Defaunation Shadow," *PNAS* 114 (2017): 8655–8659; Müller and Werner, "Urban Biodiversity"; M. T. J. Johnson and J. Munshi-South, "Evolution of Life in Urban Environments," *Science* 358 (2017): eaam8327; Lewis Mumford, "The Natural History of Urbanization," in *Man's Role in Changing the Face of the Earth,* ed. William L. Thomas Jr. (Chicago: University of Chicago Press, 1956), 382–400; William Cronon, *Nature's Metropolis* (New York: W. W. Norton, 1991); Rutherford H. Platt, "The Ecological City: Introduction and Overview," in *The Ecological City: Preserving and Restoring Urban Biodiversity,* ed. Platt, Rowan A. Rowntree, and Pamela C. Muick (Amherst: University of Massachusetts Press, 1994), 1–17; Nancy B. Grimm et al., "Global Change and the Ecology of Cities," *Science* 319 (2008): 756–760.

4. Karen C. Seto et al., "Sustainability in an Urbanizing Planet," *PNAS* 114 (2017): 8935–8938; Müller and Werner, "Urban Biodiversity"; James Corner, "The Ecological Imagination: Life in the City and the Public Realm," in *Nature and Cities: The Ecological Imperative in Urban Design and Planning,* ed. Frederick R. Steiner, George F. Thompson, and Armando Carbonell (Cambridge, Mass.: Lincoln Institute of Land Policy, 2016), 24; George F. Thompson, Frederick R. Steiner, and Armando Carbonell, "The Landscape Today and the Challenges Ahead," in Steiner, Thompson, and Carbonell, *Nature and Cities,* xxi; Kongjian Yu, "Creating Deep Forms in Urban Nature: The Peasant's Approach to Urban Design," in Steiner, Thompson, and Carbonell, *Nature and Cities,* 94–117; Richard Weller, "The City Is Not an Egg: Western Urbanization in Relation to Changing Conceptions of Nature," in Steiner, Thompson, and Carbonell, *Nature and Cities,* 48 (emphasis in original).

5. "Green space" is defined differently according to discipline and context; I define it very broadly here as "a dedicated public or private place with plants." Lucy Taylor and Dieter F. Hochuli, "Defining Greenspace: Multiple Uses across Multiple Disciplines," *Landscape and Urban Planning* 158 (2017): 25–38; Ashlea J. Hunter and Gary W. Luck, "Defining and Measuring the Social-Ecological Quality of Urban Greenspace: A Semi-Systematic Review," *Urban Ecosystems* 18 (2015): 1139–1163; Nilon et al., "Planning for the Future." Note that in Europe, "green infrastructure" is not limited to cities but is used to discuss natural areas at multiple scales, even continental: see, e.g., Tord Snäll et al., "Green Infrastructure Design Based on Spa-

tial Conservation Prioritization and Modeling of Biodiversity Features and Ecosystem Services," *Environmental Management* 57 (2016): 251–256; Mark A. Benedict and Edward T. McMahon, "Green Infrastructure: Smart Conservation for the 21st Century," *Renewable Resources Journal* 20 (2002): 12–17; Konstantinos Tzoulas et al., "Promoting Ecosystem and Human Health in Urban Areas Using Green Infrastructure: A Literature Review," *Landscape and Urban Planning* 81 (2007): 167–178; Eneko Garmendia et al., "Biodiversity and Green Infrastructure in Europe: Boundary Object or Ecological Trap?" *Land Use Policy* 56 (2016): 315–319; Ryan A. McManamay et al., "U.S. Cities Can Manage National Hydrology and Biodiversity Using Local Infrastructure Policy," *PNAS* 114 (2017): 9581–9586; Myla F. J. Aronson et al., "Biodiversity in the City: Key Challenges for Urban Green Space Management," *Frontiers in Ecology and the Environment* 15 (2017): 189–196; Christopher A. Lepczyk et al., "Biodiversity in the City: Fundamental Questions for Understanding the Ecology of Urban Green Spaces for Biodiversity Conservation," *BioScience* 67 (2017): 799–807; and Pincetl, Watt, and Santos, "Land Use Regulation."

6. Mcdonald, Kareiva, and Forman, "Implications of Current and Future Urbanization"; Christopher A. Lepczyk et al., "Global Patterns and Drivers of Urban Bird Diversity," in Murgui and Hedblom, *Ecology and Conservation of Birds*, 13–34; Karen C. Seto, Burak Güneralp, and Lucy R. Hutyra, "Global Forecasts of Urban Expansion to 2030 and Direct Impacts on Biodiversity and Carbon Pools," *PNAS* 109 (2012): 16083–16088.

7. Luck, "Review of Relationships"; James R. Miller and Richard J. Hobbs, "Conservation Where People Live and Work," *CB* 16 (2002): 330–337; Kylie Soanes and Pia E. Lentini, "When Cities Are the Last Chance for Saving Species," *Frontiers in Ecology and Environment* 17 (2019): 225–231; Aronson et al., "Global Analysis"; Aronson et al., "Biodiversity in the City"; Jack Ahern, "Urban Landscape Sustainability and Resilience: The Promise and Challenges of Integrating Ecology with Urban Planning and Design," *Landscape Ecology* 28 (2013): 1203–1212; Kylie Soanes et al., "Correcting Common Misconceptions to Inspire Conservation Action in Urban Environments," *CB* 33 (2019): 300–306; Mark J. McDonnell and Amy K. Hahs, "Adaptation and Adaptedness of Organisms to Urban Environments," *Annual Review of Ecology, Evolution, and Systematics* 46 (2015): 261–280; Aronson et al., "Global Analysis"; Ives et al., "Cities Are Hotspots"; Lepczyk et al., "Global Patterns"; Lepczyk et al., "Biodiversity."

8. Richard A. Fuller et al., "Environment and Biodiversity," in *Dimensions of the Sustainable City*, ed. Mike Jenks and Colin A. Jones (New York: Springer, 2010), 75–103; José Almiñana and Carol Franklin, "Creative Fitting: Toward Designing the

City as Nature," in Steiner, Thompson, and Carbonell, *Nature and Cities*, 154. Good design can accommodate fire risk as well: Thomas Curwen, "California's Deadliest Wildfires Were Decades in the Making: 'We Have Forgotten What We Need to Do to Prevent It,'" *LAT*, Oct. 22, 2017; Doug Smith and Nina Agrawal, "550,000 Homes in Southern California Have the Highest Risk of Fire Damage, but They Are Not Alone," *LAT,* Nov. 13, 2017; Peter Werner and Rudolf Zahner, "Urban Patterns and Biological Diversity: A Review," in Müller, Werner, and Kelcey, *Urban Biodiversity and Design,* 145–173; Lepczyk et al., "Biodiversity"; Aronson et al., "Biodiversity in the City." For the additional importance of habitat and environmental heterogeneity, see Jenny A. Hodgson et al., "Climate Change, Connectivity and Conservation Decision Making: Back to Basics," *Journal of Applied Ecology* 46 (2009): 964–969.

9. Lenore Fahrig, "Relative Effects of Habitat Loss and Fragmentation on Population Extinction," *Journal of Wildlife Management* 61 (1997): 603–610; Lenore Fahrig, "Effects of Habitat Fragmentation on Biodiversity," *Annual Review of Ecology, Evolution, and Systematics* 34 (2003): 487–515; Lenore Fahrig, "Ecological Responses to Habitat Fragmentation Per Se," *Annual Review of Ecology, Evolution, and Systematics* 48 (2017): 1–23.

10. David Lindenmayer, "Small Patches Make Critical Contributions to Biodiversity Conservation," *PNAS* 116 (2019): 717–719; Brendan A. Wintle et al., "Global Synthesis of Conservation Studies Reveals the Importance of Small Habitat Patches for Biodiversity," *PNAS* 116 (2019): 909–914; Ayesha I. T. Tulloch et al., "Understanding the Importance of Small Patches of Habitat for Conservation," *Journal of Applied Ecology* 53 (2016): 418–429; Robert F. Baldwin and Nakisha T. Fouch, "Understanding the Biodiversity Contributions of Small Protected Areas Presents Many Challenges," *Land* 7 (2019): https://doi.org/10.3390/land7040123; Lindenmayer, "Small Patches"; Fahrig, "Ecological Responses"; Soanes et al., "Correcting Common Misconceptions." The effects of artificial light haven't been well studied but could be significant: McDonnell and Hahs, "Adaptation and Adaptedness"; Lepczyk et al., "Biodiversity"; James R. Miller and Richard J. Hobbs, "Conservation Where People Live and Work," *CB* 16 (2002): 330–337; McDonnell and Hahs, "Adaptation and Adaptedness"; Jeffrey A. Brown et al., "Evaluating the Long-Term Effectiveness of Terrestrial Protected Areas: A 40-Year Look at Forest Bird Diversity," *Biodiversity and Conservation* 28 (2019): 811–826; Luis Mata et al., "Punching above Their Weight: The Ecological and Social Benefits of Pop-Up Parks," *Frontiers in Ecology and the Environment* 17 (2019): 341–347.

11. California Department of Fish and Game, *Southern California Coastal Sage Scrub NCCP Conservation Guidelines* (Sacramento: California Resources Agency,

August 1993); USFWS, "Reinitiation of Formal Consultation on Implementation of the Special Rule for the Coastal California Gnatcatcher (1-6-93-FW-37RI)" (Report prepared for USFWS, California State Supervisor, Sacramento, 1996), 14; Kenneth L. Weaver, "Coastal Sage Scrub Variations of San Diego County and Their Influence on the Distribution of the California Gnatcatcher," *WB* 29 (1998): 392–405; Clark S. Winchell and Paul F. Doherty Jr., "Using California Gnatcatcher to Test Underlying Models in Habitat Conservation Plans," *Journal of Wildlife Management* 72 (2008): 1322–1327.

12. Garmendia et al., "Biodiversity and Green Infrastructure"; Deborah Rudnick et al., *The Role of Landscape Connectivity in Planning and Implementing Conservation and Restoration Priorities*, Issues in Ecology Report No. 16 (Washington, D.C.: Ecological Society of America, 2012); David M. Theobald et al., "Connecting Natural Landscapes Using a Landscape Permeability Model to Prioritize Conservation Activities in the United States," *Conservation Letters* 5 (2012): 123–133.

13. Simon G. Dures and Graeme S. Cumming, "The Confounding Influence of Homogenising Invasive Species in a Globally Endangered and Largely Urban Biome: Does Habitat Quality Dominate Avian Biodiversity?" *BC* 142 (2010): 768–777; Trina Rytwinski et al., "How Effective Is Road Mitigation at Reducing Road-Kill? A Meta-Analysis," *PLoS ONE* 11 (2016): https://doi.org/10.1371/journal.pone.0166941; Travis Longcore et al., "Wildlife Underpass Use and Environmental Impact Assessment: A Southern California Case Study," *Cities and the Environment* 11 (2011): art4; Katherine Gammon, "Los Angeles to Build World's Largest Wildlife Bridge across 10-Lane Freeway," *Guardian,* Aug. 21, 2019; Martin Wickol, "Mountain Lions Have Good Day in Court; Tentative Ruling Questions Temecula Environmental Report," *Press-Enterprise* (Riverside, Calif.), Nov. 22, 2019; Charlie Landry, "Wildlife Crossings and Connectivity within the Western Riverside County MSHCP," presentation at the 2019 National HCP Coalition annual meeting in Shepherdstown, W. Va., Nov. 13–15, 2019, https://www.nhcpcoalition.org/2019-annual-meeting/; Robert L. Chianese, "Suburban Stalkers: The Near-Wild Lions in Our Midst," *American Scientist* 105 (2017): 278–281; T. Winston Vickers et al., "Survival and Mortality of Pumas (*Puma concolor*) in a Fragmented, Urbanizing Landscape," *PLoS ONE* 10 (2015): https://doi.org/10.1371/journal.pone.0131490.

14. Roarke Donnelly and John M. Marzluff, "Relative Importance of Habitat Quantity, Structure, and Spatial Pattern to Birds in Urbanizing Environments," *Urban Ecosystems* 9 (2006): 99–117; Dearborn and Kark, "Motivations"; Garmendia et al., "Biodiversity and Green Infrastructure"; Daniel M. Evans et al., "Species Recovery in the United States: Increasing the Effectiveness of the Endangered Species

Act," *Issues in Ecology* 20 (2016): 1–28; Patrick L. Thompson and Andrew Gonzalez, "Dispersal Governs the Reorganization of Ecological Networks under Environmental Change," *Nature Ecology and Evolution* 1 (2017): https://doi.org/10.1038/s41559-017 -0162; Marc A. Linderman and Christopher A. Lepczyk, "Vegetation Dynamics and Human Settlement across the Conterminous United States," *Journal of Maps* 9 (2013): 198–202; Caitlin E. Littlefield et al., "Connectivity for Species on the Move: Supporting Climate-Driven Range Shifts," *Frontiers in Ecology and the Environment* 17 (2019): 270–278. For an argument against overemphasis on connectivity for conservation during climate change, see Hodgson et al., "Climate Change"; Johnson and Munshi-South, "Evolution of Life"; and Lepczyk et al., "Biodiversity." There is a limit to possible elevation shifts: Mark C. Urban, "Escalator to Extinction," *PNAS* 115 (2018): 11871–11873.

15. Amanda D. Rodewald, "The Importance of Land Uses within the Landscape Matrix," *Wildlife Society Bulletin* 31 (2003): 586–592; Jerry F. Franklin and D. B. Lindenmayer, "Importance of Matrix Habitats in Maintaining Biological Diversity," *PNAS* 106 (2009): 349–350; Werner and Zahner, "Urban Patterns"; John Kostyack et al., "Beyond Reserves and Corridors: Policy Solutions to Facilitate the Movement of Plants and Animals in a Changing Climate," *BioScience* 61 (2011): 713–719; Don A. Driscoll et al., "Conceptual Domain of the Matrix in Fragmented Landscapes," *Trends in Ecology and Evolution* 28 (2013): 605–613; Ashlea J. Zivanovic and Gary W. Luck, "Social and Environmental Factors Drive Variation in Plant and Bird Communities across Urban Greenspace in Sydney, Australia," *Journal of Environmental Management* 169 (2016): 210–222; Soanes et al., "Correcting Common Misconceptions"; Laura R. Prugh et al., "Effects of Habitat Area and Isolation on Fragmented Animal Populations," *PNAS* 105 (2008): 20770–20775; J. Amy Belaire, Christopher J. Whelan, and Emily S. Minor, "Having Our Yards and Sharing Them Too: The Collective Effects of Yards on Native Bird Species in an Urban Landscape," *Ecological Applications* 24 (2014): 2132–2143; Derric N. Pennington, James Hansel, and Robert B. Blair, "The Conservation Value of Urban Riparian Areas for Landbirds during Spring Migration: Land Cover, Scale, and Vegetation Effects," *BC* 141 (2008): 1235–1248; Chad L. Seewageh, Eric J. Slayton, and Christopher G. Guglielmo, "Passerine Migrant Stopover Duration and Spatial Behavior at an Urban Stopover Site," *Acta Oecologica* 36 (2010): 484–492; Amanda J. Oliver et al., "Avifauna Richness Enhanced in Large, Isolated Urban Parks," *Landscape and Urban Planning* 102 (2011): 215–225; Sarah E. Reed et al., *Wildlife Response to Human Recreation on NCCP Reserves in San Diego County,* Wildlife Conservation Society, Final Report, Agreement No. P1182112 (https://nrm.dfg.ca.gov/FileHandler.ashx?DocumentID=99222, 2014); Courtney L.

Larson et al., "Accessibility Drives Species Exposure to Recreation in Fragmented Urban Reserve Network," *Landscape and Urban Planning* 175 (2018): 62–71.

16. Franklin and Lindenmayer, "Importance of Matrix Habitats"; Dures and Cumming, "Confounding Influence"; California Department of Fish and Wildlife, *Southern California Coastal Sage Scrub NCCP Conservation Guidelines;* Prugh et al., "Effects of Habitat Area"; James I. Watling et al., "Meta-Analysis Reveals the Importance of Matrix Composition for Animals in Fragmented Habitat," *Global Ecology and Biogeography* 20 (2011): 209–217; Michael W. Strohbach, Susannah B. Lerman, and Paige S. Warren, "Are Small Greening Areas Enhancing Bird Diversity? Insights from Community-Driven Greening Projects in Boston," *Landscape and Urban Planning* 114 (2013): 69–79. The matrix may not contribute as much benefit as habitat patches and quality to some taxonomic groups: Jayme A. Prevedello and Marcus V. Vieira, "Does the Type of Matrix Matter? A Quantitative Review of the Evidence," *Biodiversity and Conservation* 19 (2010): 1205–1223.

17. Belaire, Whelan, and Minor, "Having Our Yards"; Katherine N. Irvine et al., "Ecological and Psychological Value of Urban Green Space," in Jenks and Jones, *Dimensions of the Sustainable City,* 215–237; Dustin R. Partridge and J. Alan Clark, "Urban Green Roofs Provide Habitat for Migrating and Breeding Birds and Their Arthropod Prey," *PLoS ONE* 13 (2018): https://doi.org/10/1371/journal.pone.0202298. Green roofs on very tall buildings may be detrimental to some species: see Lepczyk et al., "Biodiversity"; Nate Millington, "From Urban Scar to 'Park in the Sky': *Terrain Vague,* Urban Design, and the Remaking of New York City's High Line Park," *Environmental Planning A* 47 (2015): 2324–2338; Paul H. Gobster, Sonya Sachdeva, and Greg Lindsey, "Up on the 606: Understanding the Use of a New Elevated Pedestrian and Bicycle Trail in Chicago, Illinois," *Transportation Research Record* 2644 (2017): 83–91; Alessandro Rigolon and Jeremy Németh, "'We're Not in the Business of Housing': Environmental Gentrification and the Nonprofitization of Green Infrastructure Projects," *Cities* 81 (2018): 71–80; Weiqi Zhou, Brendan Fisher, and Stewart T. A. Pickett, "Cities Are Hungry for Actionable Ecological Knowledge," *Frontiers in Ecology and the Environment* 17 (2019): 135; David B. Lindenmayer and Joern Fischer, *Habitat Fragmentation and Landscape Change: An Ecological and Conservation Synthesis* (Washington, D.C.: Island Press, 2006); Masashi Soga et al., "Land Sharing vs. Land Sparing: Does the Compact City Reconcile Urban Development and Biodiversity Conservation?" *Journal of Applied Ecology* 51 (2014): 1378–1386; Lydia Collas et al., "Urban Development, Land Sharing and Land Sparing: The Importance of Considering Restoration," *Journal of Applied Ecology* 54 (2017): 1865–1873; Fiona M. Caryl et al., "Functional Responses of Insectivorous Bats to Increasing Housing Den-

sity Support 'Land-Sparing' Rather than 'Land-Sharing' Urban Growth Strategies," *Journal of Applied Ecology* 53 (2016): 191–201; Andrew Geschke et al., "Compact Cities or Sprawling Suburbs? Optimising the Distribution of People in Cities to Maximise Species Diversity," *Journal of Applied Ecology* 55 (2018): 2320–2331; Soanes et al., "Correcting Common Misconceptions"; Fahrig, "Ecological Responses"; Fuller et al., "Environment and Biodiversity"; Irvine et al., "Ecological and Psychological Value"; Aronson et al., "Biodiversity in the City"; Iain Stott et al., "Land Sparing Is Crucial for Urban Ecosystem Services," *Frontiers in Ecology and the Environment* 13 (2015): 387–393; Caryl et al., "Functional Responses"; Nélida R. Villaseñor et al., "Compact Development Minimizes the Impacts of Urban Growth on Native Mammals," *Journal of Applied Ecology* 54 (2017): 794–804.

18. Benedict and McMahon, "Green Infrastructure"; Irvine et al., "Ecological and Psychological Value"; Oliver et al., "Avifaunal Richness"; Aronson et al., "Biodiversity in the City"; Mark A. Goddard, Karen Ikin, and Susannah B. Lerman, "Ecological and Social Factors Determining the Diversity of Birds in Residential Yards and Gardens," in *Ecology and Conservation of Birds, in Urban Environments,* ed. Enrique Murgui and Marcus Hedblom (Cham, Switzerland: Springer, 2017), 371–398.

19. Dures and Cumming, "Confounding Influence"; Belaire, Whelan, and Minor, "Having Our Yards"; D. T. T. Flockhart, D. R. Norris, and J. B. Coe, "Predicting Free-Roaming Cat Population Densities in Urban Areas," *Animal Conservation* 19 (2016): 472–483; Catherine M. Hall et al., "Factors Determining the Home Ranges of Pet Cats: A Meta-Analysis," *BC* 203 (2016): 313–320; Goddard, Ikin, and Lerman, "Ecological and Social Factors"; Dagny Krauze-Gryz, Michał Żmihorski, and Jakub Gryz, "Annual Variation in Prey Composition of Domestic Cats in Rural and Urban Environment," *Urban Ecosystems* 20 (2017): 945–952; Kayleigh Chalkowski et al., "Who Let the Cats Out? A Global Meta-Analysis on Risk of Parasitic Infection in Indoor versus Outdoor Domestic Cats (*Felis catus*)," *Biology Letters* 15 (2019): 20180840; Scott R. Loss, Tom Will, and Peter P. Marra, "The Impact of Free-Ranging Domestic Cats on Wildlife of the United States," *Nature Communications* 4 (2013): art1396; Brett P. Murphy et al., "Introduced Cats (*Felis catus*) Eating a Continental Fauna: The Number of Mammals Killed in Australia," *BC* 237 (2019): 28–40; Kerrie Anne T. Loyd, Sonia M. Hernandez, and David L. McRuer, "The Role of Domestic Cats in the Admission of Injured Wildlife at Rehabilitation and Rescue Centers," *Wildlife Society Bulletin* 41 (2017): 55–61; P. J. Baker, R. Thompson, and A. Grogan, "Survival Rates of Cat-Attacked Birds Admitted to RSPCA Wildlife Centres in the UK: Implications for Cat Owners and Wildlife Rehabilitators," *Animal Welfare* 27 (2018): 305–318. Cat owners in the United States were particularly unconvinced by

the risk to wildlife that cats posed; just 20 percent of U.S. owners thought this was a serious problem: Catherine M. Hall et al., "Community Attitudes and Practices of Urban Residents Regarding Predation by Pet Cats on Wildlife: An International Comparison," *PLoS ONE* 11, no. 4 (2016): e0151962; Christopher A. Lepczyk et al., "A Review of Cat Behavior in Relation to Disease Risk and Management Options," *Applied Animal Behaviour Science* 173 (2015): 29–39.

20. Hill, "Form Follows Flows"; Nancy B. Grimm et al., "Integrated Approaches to Long-Term Studies of Urban Ecological Systems," *BioScience* 50 (2000): 571–584; Marina Alberti et al., "Integrating Humans into Ecology: Opportunities and Challenges for Studying Urban Ecosystems," *BioScience* 53 (2003): 1169–1179; Pataki et al., "Urban Ecosystems." See chapters by Forster Ndubisi, Charles Waldheim, Richard Weller, and Kongjian Yu in Steiner, Thompson, and Carbonell, *Nature and Cities;* Nilon et al., "Planning for the Future"; Spirn, *Granite Garden: Urban Nature and Human Design* (New York: Basic Books, 1984); Dearborn and Kark, "Motivations"; Fuller et al., "Environment and Biodiversity"; Hannah Hoag, "How Cities Can Beat the Heat," *Nature* 524 (2015): 402–404; Peleg Kremer, Zoé A. Hamstead, and Timon McPhearson, "The Value of Urban Ecosystem Services in New York City: A Spatially Explicit Multicriteria Analysis of Landscape Scale Valuation Scenarios," *Environmental Science and Policy* 62 (2016): 57–68; S. J. Livesley, E. G. McPherson, and C. Calfapletra, "The Urban Forest and Ecosystem Services: Impacts on Urban Water, Heat, and Pollution Cycles at the Tree, Street, and City Scale," *Journal of Environmental Quality* 45 (2016): 119–124; Anne Whiston Spirn, "The Granite Garden: Where Do We Stand Today?" in Steiner, Thompson, and Carbonell, *Nature and Cities,* 51–68; Katherine J. Willis and Gillian Petrokofsky, "The Natural Capital of City Trees," *Science* 356 (2017): 374–376; Carly D. Ziter et al., "Scale-Dependent Interactions between Tree Canopy Cover and Impervious Surfaces Reduce Daytime Urban Heat during Summer," *PNAS* 116 (2019): 7575–7580; Benedict and McMahon, "Green Infrastructure"; Joshua P. Newell et al., "Green Alley Programs: Planning for a Sustainable Urban Infrastructure?" *Cities* 31 (2013): 144–155; Katharine R. E. Sims et al., "Assessing the Local Economic Impacts of Land Protection," *CB* 33 (2019): 1035–1044.

21. Jeffrey E. Zabel and Robert W. Paterson, "The Effects of Critical Habitat Designation on Housing Supply: An Analysis of California Housing Construction Activity," *Journal of Regulatory Science* 46 (2006): 67–95; Stephen Polasky et al., "Where to Put Things? Spatial Land Management to Sustain Biodiversity and Economic Returns," *BC* 141 (2008): 1505–1524; Harvey Molotch, "The City as Growth Machine: Toward a Political Economy of Place," *American Journal of Sociology* 8 (1976): 309–

332; Benedict and McMahon, "Green Infrastructure"; E. Gregory McPherson et al., "The Structure, Function, and Value of Urban Forests in California Communities," *Urban Forestry and Urban Greening* 28 (2017): 43–53; Fuller et al., "Environment and Biodiversity"; Katherine J. Turo and Mary M. Gardiner, "From Potential to Practical: Conserving Bees in Urban Public Green Spaces," *Frontiers in Ecology and the Environment* 17 (2019): 167–175; Dearborn and Kark, "Motivations"; Garmendia et al., "Biodiversity and Green Infrastructure"; Lepczyk et al., "Biodiversity"; Seto et al., "Sustainability"; Miller and Hobbs, "Conservation Where People Live"; Press, *Saving Open Space,* 135; Luck, "Review of Relationships."

22. Melissa Checker, "Wiped Out by the 'Greenwave': Environmental Gentrification and the Paradoxical Politics of Urban Sustainability," *City and Society* 23 (2011): 210–229; Jennifer R. Wolch, Jason Byrne, and Joshua P. Newell, "Urban Green Space, Public Health, and Environmental Justice: The Challenge of Making Cities 'Just Green Enough,'" *Landscape and Urban Planning* 125 (2014): 234–244; Ronald Ngom, Pierre Gosselin, and Claudia Blais, "Reduction of Disparities in Access to Green Spaces: Their Geographic Insertion and Recreation Functions Matter," *Applied Geography* 66 (2016): 35–51; Steven Lang and Julia Rothenberg, "Neoliberal Urbanism, Public Space, and the Greening of the Growth Machine: New York City's High Line Park," *Environment and Planning A* 49 (2017): 1743–1761. The "luxury effect" is a converse relationship: see Aronson et al., "Biodiversity in the City"; and Misha Leong, Robert R. Dunn, and Michelle D. Trautwein, "Biodiversity and Socioeconomics in the City: A Review of the Luxury Effect," *Biology Letters* 14 (2018): https://doi.org/10.1098/rsbl.2018.0082; Rolf Pendall et al., "Connecting Smart Growth, Housing Affordability, and Racial Equity," in *The Geography of Opportunity: Race and Housing Choice,* ed. Xavier de Souza Briggs (Washington, D.C.: Brookings Institution Press, 2005); Matthew Jerzyk, "Gentrification's Third Way: An Analysis of Housing Policy and Gentrification in Providence," *Harvard Law and Policy Review* 3 (2009): 413–430; Galen Cranz, *Politics of Park Design: A History of Urban Parks in America* (Cambridge, Mass.: MIT Press, 1982); Kevin Loughran, "Parks for Profit: The High Line, Growth Machines, and the Uneven Development of Urban Public Spaces," *City and Community* 13 (2014): 49–68; Parama Roy, "Collaborative Planning—a Neoliberal Strategy? A Study of the Atlanta BeltLine," *Cities* 43 (2015): 59–68; Kenneth A. Gould and Tammy L. Lewis, *Green Gentrification: Urban Sustainability and the Struggle for Environmental Justice* (New York: Routledge, 2017); Dan Immergluck and Tharunya Balan, "Sustainable for Whom? Green Urban Development, Environmental Gentrification, and the Atlanta BeltLine," *Urban Geography* 39 (2017): 546–562; Darren J.

Patrick, "The Matter of Displacement: A Queer Urban Ecology of New York City's High Line," *Social and Cultural Geography* 15 (2014): 920-941.

23. Millington, "Urban Scar"; Rigolon and Németh, "'Not in the Business of Housing'"; Wolch, Byrne, and Newell, "Urban Green Space"; Robert J. Sampson, "Urban Sustainability in an Age of Enduring Inequalities: Advancing Theory and Ecometrics for the 21st-Century City," *PNAS* 114 (2017): 8957-8962; Laura Grant and Christian Langpap, "Private Provision of Public Goods by Environmental Groups," *PNAS* 116 (2019): 5334-5340; Anna Jorgensen and Paul H. Gobster, "Shades of Green: Measuring the Ecology and Urban Green Space in the Context of Human Health and Well-Being," *Nature and Culture* 5 (2010): 338-363; Hunter and Luck, "Defining and Measuring."

24. Masashi Soga and Kevin J. Gaston, "Extinction of Experience: The Loss of Human-Nature Interactions," *Frontiers in Ecology and the Environment* 14 (2016): 94-101; Masashi Soga et al., "Reducing the Extinction of Experience: Association between Urban Form and Recreational Use of Public Greenspace," *Landscape and Urban Planning* 143 (2015): 69-75; Caoimhe Twohig-Bennett and Andy Jones, "The Health Benefits of the Great Outdoors: A Systematic Review and Meta-Analysis of Greenspace Exposure and Health Outcomes," *Environmental Research* 166 (2018): 628-637; F. Law Olmsted, "The Yosemite Valley and the Mariposa Big Tree: A Preliminary Report" (1865), https://www.nps.gov/parkhistory/online_books/anps/anps_1b.htm; Ethan Carr, "Geniuses of Place," *Nature* 535 (2016): 34-36; Riikka Puhakka, Kati Pitkänen, and Pirkko Siikamäki, "The Health and Well-Being Impacts of Protected Areas in Finland," *Journal of Sustainable Tourism* 25 (2017): 1830-1847; Julien Terraube, Álvaro Fernández-Llamazares, and Mar Cabeza, "The Role of Protected Areas in Supporting Human Health: A Call to Broaden the Assessment of Conservation Outcomes," *Current Opinion in Environmental Sustainability* 25 (2017): 50-58; Alexandra Jiricka-Pürrer et al., "Do Protected Areas Contribute to Health and Well-Being? A Cross-Cultural Comparison," *International Journal of Environmental Research and Public Health* 16 (2019): http://doi.org/10.3390/ijerph16071172; Dearborn and Kark, "Motivations."

25. Irvine et al., "Ecological and Psychological Value"; Danielle F. Shanahan et al., "The Health Benefits of Urban Nature: How Much Do We Need?" *BioScience* 65 (2015): 476-485; Wolch, Byrne, and Newell, "Urban Green Space"; Timothy Beatley, "New Directions in Urban Nature: The Power and Promise of Biophilic Cities and Blue Urbanism," in Steiner, Thompson, and Carbonell, *Nature and Cities,* 265-285; Twohig-Bennett and Jones, "Health Benefits"; Catherine Ward Thompson et al.,

"More Green Space Is Linked to Less Stress in Deprived Communities: Evidence from Salivary Cortisol Patterns," *Landscape and Urban Planning* 105 (2019): 221–229. A few of these papers also identify some risks that green space may represent to urban dwellers, such as increased disease reservoirs for mosquito-borne illnesses. Peter P. Groenewegen et al., "Vitamin G: Effects of Green Space on Health, Well-Being, and Social Safety," *BMC Public Health* 6 (2006): https://doi.org/10.1186/1471-2458 -6-149; Laura E. Jackson, "The Relationship of Urban Design to Human Health and Condition," *Landscape and Urban Planning* 64 (2003): 191–200; Tzoulas et al., "Promoting Ecosystem and Human Health"; Pataki et al., "Urban Ecosystems"; Forest Isbell et al., "Linking the Influence and Dependence of People on Biodiversity across Scales," *Nature* 546 (2017): 65–72; Twohig-Bennett and Jones, "Health Benefits"; Omid Kardan et al., "Neighborhood Green-Space and Health in a Large Urban Center," *Scientific Reports* 5 (2015): https://doi.org/10.1038/srep11610; Qing Li, "Effect of Forest Bathing Trips on Human Immune Function," *Environmental Health and Preventative Medicine* 15 (2010): 9–17; Chorong Song, Harumi Ikei, and Yoshifumi Miyazaki, "Physiological Effects of Nature Therapy: A Review of the Research in Japan," *International Journal of Environmental Research and Public Health* 13 (2016): https://doi.org/10.3390/ijerph13080781; Margaret M. Hansen, Reo Jones, and Kirsten Tocchini, "Shinrin-Yoku (Forest Bathing) and Nature Therapy: A State-of-the-Art Review," *International Journal of Environmental Research and Public Health* 14 (2017): https://doi.org/10.3390/ijerph14080851; Qing Li et al., "Acute Effects of Walking in Forest Environments on Cardiovascular and Metabolic Parameters," *European Journal of Applied Physiology* 111 (2011): 2845–2853; Sjerp de Vries et al., "Natural Environments—Healthy Environments? An Exploratory Analysis of the Relationship between Greenspace and Health," *Environment and Planning A* 35 (2003): 1717–1731; Frances E. Kuo and William C. Sullivan, "Aggression and Violence in the Inner City: Effects of Environment via Mental Fatigue," *Environment and Behavior* 33 (2001): 543–571; Frances E. Kuo and William C. Sullivan, "Environmental and Crime in the Inner City: Does Vegetation Reduce Crime?" *Environment and Behavior* 33 (2001): 343–367; Mary K. Wolfe and Jeremy Mennis, "Does Vegetation Encourage or Suppress Urban Crime? Evidence from Philadelphia, PA," *Landscape and Urban Planning* 108 (2012): 112–122; Sandra Bogar and Kirsten M. Beyer, "Green Space, Violence, and Crime: A Systematic Review," *Trauma, Violence, and Abuse* 17 (2016): 160–171; John S. Ji et al., "Residential Greenness and Mortality in Oldest-Old Women and Men in China: A Longitudinal Cohort Study," *Lancet Planet Health* 3 (2019): https://doi.org/10.1016/S2542-5196(18)30264-X; Austin Troy, J. Morgan Grove, and Jarlath O'Neil-Dunne, "The Relationship between Tree Canopy and Crime Rates across an

Urban-Rural Gradient in the Greater Baltimore Region," *Landscape and Urban Planning* 106 (2012): 262–270; Michelle C. Kondo et al., "The Association between Urban Trees and Crime: Evidence from the Spread of the Emerald Ash Borer in Cincinnati," *Landscape and Urban Planning* 157 (2017): 193–199.

26. Richard Louv, *Last Child in the Woods: Saving Our Children from Nature-Deficit Disorder* (New York: Algonquin Books, 2008); George Monbiot, *Feral: Rewilding the Land, the Sea, and Human Life* (Chicago: University of Chicago Press, 2014); Peter H. Kahn Jr. and Stephen R. Kellert, eds., *Children and Nature: Psychological, Sociocultural, and Evolutionary Investigations* (Cambridge, Mass.: MIT Press, 2002); Payam Dadvand et al., "Green Spaces and Cognitive Development in Primary Schoolchildren," *PNAS* 112 (2015): 7937–7942; Soga and Gaston, "Extinction of Experience"; Kristine Engemann et al., "Residential Green Space in Childhood Is Associated with Lower Risk of Psychiatric Disorders from Adolescence into Adulthood," *PNAS* 116 (2019): 5188–5193; Graham A. Rook, "Regulation of the Immune System by Biodiversity from the Natural Environment: An Ecosystem Service Essential to Health," *PNAS* 110 (2013): 18360–18367; Terry Hartig et al., "Nature and Health," *Annual Review of Public Health* 35 (2014): 207–228; Peter James et al., "A Review of the Health Benefits of Greenness," *Current Epidemiology Reports* 2 (2015): 131–142; Twohig-Bennett and Jones, "Health Benefits."

27. National League of Cities website: https://www.nlc.org; City of Austin, TX, Children in Nature Collaborative website: http://www.austintexas.gov/sites/default/files/files/COBOR_Resolution.pdf.

28. Edward O. Wilson, *Biophilia: The Human Bond with Other Species* (Cambridge, Mass.: Harvard University Press, 1984); Robert M. Pyle, *The Thunder Tree: Lessons from an Urban Wildland* (Boston: Houghton Mifflin, 1993); Nancy M. Wells and Kristi S. Lekies, "Nature and the Life Course: Pathways from Childhood Nature Experiences to Adult Environmentalism," *Children, Youth, and Environments* 16 (2006): 1–24; Catharine Ward Thompson, Peter Aspinall, and Alicia Montarzino, "The Childhood Factor: Adult Visits to Green Places and the Significance of Childhood Experience," *Environmental Behaviors* 40 (2008): 111–143; Soga et al., "Reducing Extinction of Experience"; Beatley, "New Directions"; Timothy Beatley, *Biophilic Cities: Integrating Nature into Urban Design and Planning* (Washington, D.C.: Island Press, 2010); Netta Weinstein, Andrew K. Przybylski, and Richard M. Ryan, "Can Nature Make Us More Caring? Effects of Immersion in Nature on Intrinsic Aspirations and Generosity," *Personality and Social Psychology Bulletin* 35 (2009): 1315–1329; Arianne J. Van der Wal et al., "Do Natural Landscapes Reduce Future Discounting in Humans?" *PRSB* 280 (2013): https://doi.org/10.1098/rspb.2013.2295.

29. Michael L. Rosenzweig, *Win-Win Ecology: How the Earth's Species Can Survive in the Midst of Human Enterprise* (Oxford: Oxford University Press, 2003).

30. Generally these are "strict protection" Categories I and II (sometimes also Category III) of the World Conservation Union's (IUCN) protected area classifications: see table 2 in Lisa Naughton-Treves, Margaret Buck Holland, and Katrina Brandon, "The Role of Protected Areas in Conserving Biodiversity and Sustaining Local Livelihoods," *Annual Review of Environment and Resources* 30 (2005): 219–252; United Nations Environmental Programme, "Mapping the World's Special Places," https://www.unep-wcmc.org/featured-projects/mapping-the-worlds-special-places; Robert Poirier and David Ostergren, "Evicting People from Nature: Indigenous Land Rights and National Parks in Australia, Russia, and the United States," *Natural Resources* 42 (2002): 331–351; Stan Stevens, ed., *Indigenous Peoples, National Parks, and Protected Areas: A New Paradigm Linking Conservation, Culture, and Rights* (Tucson: University of Arizona Press, 2014); Mark Dowie, *Conservation Refugees: The Hundred-Year Conflict Between Global Conservation and Native Peoples* (Cambridge, Mass.: MIT Press, 2009); John Vidal, "The Tribes Paying the Brutal Price of Conservation," *Guardian,* Aug. 28, 2016; William D. Newmark, "Legal and Biotic Boundaries of Western North American National Parks: A Problem of Congruence," *BC* 33 (1985): 197–208; Rosenzweig, *Win-Win Ecology,* 146–147; Alisa A. Wade, David M. Theobald, and Melinda J. Laituri, "A Multi-Scale Assessment of Local and Contextual Threats to Existing and Potential U.S. Protected Areas," *Landscape and Urban Planning* 101 (2011): 215–227; Andrew J. Hansen et al., "Exposure of U.S. National Parks to Land Use and Climate Change, 1900–2100," *Ecological Applications* 24 (2014): 484–502; C. Kremen and A. M. Merenlender, "Landscapes That Work for Biodiversity and People," *Science* 362 (2018): https://doi.org/10.1126/science.aau6020.

31. Bruce A. Robertson, Jennifer S. Rehage, and Andrew Sih, "Ecological Novelty and the Emergence of Evolutionary Traps," *Trends in Ecology and Evolution* 28 (2013): 552–560; Lepczyk et al., "Biodiversity"; Robin Hale and Stephen E. Swearer, "When Good Animals Love Bad Restored Habitats: How Maladaptive Habitat Selection Can Constrain Restoration," *Journal of Applied Ecology* 54 (2017): 1478–1486; Peter A. Bowler, "Ecological Restoration of Coastal Sage Scrub and Its Potential Role in Habitat Conservation Plans," *Environmental Management* 26 (2000): S85–S96; J. C. Burger et al., "Restoring Arthropod Communities in Coastal Sage Scrub," *CB* 17 (2003): 460–467; Travis Longcore, "Terrestrial Arthropods and Indicators of Ecological Restoration Success in Coastal Sage Scrub (California, U.S.A.)," *Restoration Ecology* 11 (2003): 397–409.

32. Jeffrey C. Milder, "A Framework for Understanding Conservation Develop-

ment and Its Ecological Implications," *BioScience* 57 (2007): 757–768; Karen Ikin et al., "Key Lessons for Achieving Biodiversity-Sensitive Cities and Towns," *Ecological Management and Restoration* 16 (2015): 206–214; Georgia E. Garrard et al., "Biodiversity Sensitive Urban Design," *Conservation Letters* 11 (2018): https://doi.org/10.1111/conl.12411; Millington, "Urban Scar"; Irvine et al., "Ecological and Psychological Value"; Turo and Gardener, "Potential to Practical"; Andy Millard, "Cultural Aspects of Urban Biodiversity," in Müller, Werner, and Kelcey, *Urban Biodiversity and Design*, 56–80; Joan Iverson Nassauer, "Landscape as a Medium and Method for Synthesis in Urban Ecological Design," *Landscape and Urban Planning* 106 (2012): 221–229; Alexandra Gulachenski et al., "Abandonment, Ecological Assembly and Public Health Risks in Counter-Urbanizing Cities," *Sustainability* 8 (2016): https://doi.org/10.3390/su8050491; Christine C. Rega-Brodsky, Charles H. Nilon, and Paige S. Warren, "Balancing Urban Biodiversity Needs and Resident Preferences for Vacant Lot Management," *Sustainability* 10 (2018): https://doi.org/10.3390/su10051679.

33. Benedict and McMahon, "Green Infrastructure"; Thomas L. Daniels and Mark Lapping, "Land Preservation: An Essential Ingredient in Smart Growth," *Journal of Planning Literature* 19 (2005): 316–329; Ahern, "Urban Landscape Sustainability"; Soga and Gaston, "Extinction of Experience"; Spirn, "Granite Garden"; Turo and Gardiner, "Potential to Practical"; Zhou, Fisher, and Pickett, "Cities Are Hungry"; Millington, "Urban Scar"; Rigolon and Németh, "'Not in the Business of Housing'"; James E. M. Watson et al., "Protect the Last of the Wild," *Nature* 563 (2018): 27–30; David W. Stevens, "Utility Participation in a Multispecies Plan," *Environmental Management* 20 (1996): 841–847; Andrew MacKenzie, *Understanding Metropolitan Landscapes* (New York: Routledge, 2020), 49–54.

34. Louis Sahagun, "Rare Birds Find Southern California Beach Housing," *LAT*, May 9, 2017; Dan Weikel and Louis Sahagun, "Santa Monica's New Back-to-Nature Beach Project Has Drawn the Attention of Rare Birds; But Can Beach-Goers Let Them Live in Peace?" *LAT*, May 10, 2017; Suckling et al., *Wild Success;* Morgan Greene, "Plover and Out: The Summer Stars of Montrose Beach Head South for the Winter," *Chicago Tribune*, Aug. 29, 2019.

35. Louis Sahagun, "Where Owls Roam Again; Rare Animals, Plants Are Thriving at LAX Dunes," *LAT*, Feb. 17, 2019.

36. Louis Sahagun, "Endangered Bighorn Sheep Sharing a Controversial Lush Life on the Greens at La Quita's Desert Golf Courses," *LAT*, Jan. 16, 2017.

37. Dan Flores, *Coyote America: A Natural and Supernatural History* (New York: Basic Books, 2016); Kevin R. Crooks and Michael E. Soulé, "Mesopredator Release and Avifaunal Extinctions in a Fragmented System," *Nature* 400 (1999): 563–566.

38. Chris O'Neal, "Supervisors Approve Wildlife Corridor," *VCReporter*, May 14, 2019.

39. Barry A. Sanders, "L.A.—from Park Poor to Park Rich, One Park at a Time," *LAT*, Jan. 15, 2014; Beatley, "New Directions"; Christopher Hawthorne, "What Does Building George Lucas' Museum at Exposition Park Say about L.A.?" *LAT*, Jan. 11, 2017. A redesign of Exposition Park, timed with the Lucas museum, hopes to increase the importance and usability of its green space: Sam Lubell, "Exposition Park Plans a Makeover That Would Make Seurat Smile," *LAT*, May 30, 2019.

40. Mary K. Chase et al., "Single Species as Indicators of Species Richness and Composition in California Coastal Sage Scrub Birds and Small Mammals," *CB* 14 (2000): 474–487; Rosenzweig, *Win-Win Ecology*, 88–89, 101; MacKenzie, *Understanding Metropolitan Landscapes*, 131–148.

Epilogue

1. USFWS, "Habitat Conservation Plan for the Coastal California Gnatcatcher; Categorical Exclusion for 93-129 Ltd, Orange County, California," 84 Fed. Reg. 28578 (June 19, 2019).

2. City of Laguna Niguel Planning Commission, "Time Extension for Tentative Parcel Map TPM 93-129 and Minor Use Permit UP 07-03," Jan. 24, 2017. Website for Bear Brand Ranch Community Association: http://www.progressivecm.com/bear brand/index.html.

3. USFWS, "Endangered and Threatened Species; Issuance of Enhancement of Survival and Incidental Take Permits for Safe Harbor Agreements, Candidate Conservation Agreements, Habitat Conservation Plans, and Recovery Activities, January 1, 2019, through December 31, 2019," 85 Fed. Reg. 23992 (Apr. 30, 2020).

Index

30°N latitude (ancient seaway), 38, 75,
78, 80, 83, 87
50 Parks Initiative, 192
606, the (Chicago), 181, 184

abandoned areas, 181, 183, 189
adaptive capacity, 163, 178, 256n18
adaptive management, 121, 157, 165
Administrative Procedures Act of
1946, 113, 225n24, 248n23; procedural error lawsuits, 81–82, 131
affordable housing, 168, 196; and environmental gentrification, 184;
shortage, 168, 197
agency discretion. *See* discretion
agriculture, 4, 7–8, 18, 57, 124, 133,
172; early indigenous, 4; and land
use change, 7, 18, 57, 88, 160, 172,
203n16; threat to biodiversity, 18,
57, 64, 124, 133, 177
Aichi Biodiversity Target, 172
allowable take. *See* California gnatcatcher: incidental take limits
American Ornithological Union, 76
Angeles National Forest. *See* National
Forests
Anheuser-Busch LLC. *See* Warm
Springs Ranch
Annenberg, Wallis, 175
Annenberg Foundation, 175

Artemisia. *See* California sagebrush
at-risk species, 16, 116, 173
Atwood, Jonathan "Jon": controversy,
81–82, 225n24; early work, 42–43,
75–77; funding, 85; gnatcatcher
observations, 33, 47, 49, 52; gnatcatcher subspecies, 77–80 fig. 4.1,
130; warnings, 55, 120–121
Audubon Society: California, 163;
National, 46, 102; San Bernardino
Valley, 123; San Diego, 148, 163
Australia, 18, 69, 173, 182, 201n5
Australian cities, 173, 177

Babbitt, Bruce (secretary of interior),
119, 126
Bailey, Eric, 31, 35
Baja California Peninsula (Mexico), 5,
38, 53–55, 57, 74–80, 83, 85–86, 132
Baker, Howard (U.S. senator), 108
Bald eagle (*Haliaeetus leucocephalus*),
ix, 92, 94, 106, 116
Bald Eagle Protection Act of 1940, 92
ballot initiatives: Proposition 4, 14;
Proposition 12, 136; Proposition
13 (Prop 13), 14–15, 119, 204n26;
Proposition 26, 14; Proposition 62,
14; Proposition 218, 14; Proposition
268, 14; revenue-limiting, 14–15,
119, 158–159

winter: precipitation, 2, 17, 40, 49, 61; temperatures, 36, 38, 39, 60–61
Wood, Jonathan, 100
Woods, Robert, 22, 24, 28, 35
World War II, 8, 10
World Wildlife Fund, 120

yellow-billed cuckoo (*Coccyzus americanus*), 65
Yellowstone National Park, 93, 94

Yorba family, 138
Younger Dryas. *See* climate change

Zink, Robert: California gnatcatcher species, 76, 82; California gnatcatcher subspecies, 82–83, 85, 86–87; critics of work, 85–86, 87; funding, 85, 227n30; subspecies concept, 84–85, 221n5
zoning, 11, 145, 147, 161, 191, 203n16